Annals of Mathematics Studies

Number 208

Arnold Diffusion for Smooth Systems of Two and a Half Degrees of Freedom

Vadim Kaloshin

Ke Zhang

PRINCETON UNIVERSITY PRESS

PRINCETON AND OXFORD

2020

© 2020 Princeton University Press

Requests for permission to reproduce material from this work should be sent to Permissions, Princeton University Press

Published by Princeton University Press,
41 William Street, Princeton, New Jersey 08540
6 Oxford Street, Woodstock, Oxfordshire OX20 1TR

press.princeton.edu

All Rights Reserved

Library of Congress Control Number: 2020940418
ISBN: 9780691202532
ISBN (pbk.): 9780691202525
ISBN (e-book): 9780691204932

British Library Cataloging-in-Publication Data is available

Editorial: Susannah Shoemaker and Kristen Hop
Production Editorial: Nathan Carr
Cover Design: Leslie Flis
Production: Brigid Ackerman
Publicity: Matthew Taylor and Katie Lewis
Copyeditor: Theresa Kornak

The publisher would like to acknowledge the authors of this volume for providing the camera-ready copy from which this book was printed

This book has been composed in LaTeX

Printed on acid-free paper.

Printed in the United States of America

10   9   8   7   6   5   4   3   2   1

Dedicated to the memory of John Mather,
a great mathematician and a remarkable person

Contents

Preface   xi

Acknowledgments   xiii

I   Introduction and the general scheme   1

1   Introduction   3
1.1   Statement of the result   3
1.2   Scheme of diffusion   7
1.3   Three regimes of diffusion   11
1.4   The outline of the proof   12
1.5   Discussion   14

2   Forcing relation   17
2.1   Sufficient condition for Arnold diffusion   17
2.2   Diffusion mechanisms via forcing equivalence   18
2.3   Invariance under the symplectic coordinate changes   20
2.4   Normal hyperbolicity and Aubry-Mather type   22

3   Normal forms and cohomology classes at single resonances   24
3.1   Resonant component and non-degeneracy conditions   24
3.2   Normal form   26
3.3   The resonant component   29

4   Double resonance: geometric description   31
4.1   The slow system   31
4.2   Non-degeneracy conditions for the slow system   32
4.3   Normally hyperbolic cylinders   34
4.4   Local maps and global maps   36

5   Double resonance: forcing equivalence   39
5.1   Choice of cohomologies for the slow system   39
5.2   Aubry-Mather type at a double resonance   42
5.3   Connecting to Γ_{k_1,k_2} and $\Gamma_{k_1}^{SR}$ 44
    5.3.1   Connecting to the double resonance point   44
    5.3.2   Connecting single and double resonance   45

5.4   Jump from non-simple homology to simple homology   49
5.5   Forcing equivalence at the double resonance   49

II   Forcing relation and Aubry-Mather type   53

6   Weak KAM theory and forcing equivalence   55
6.1   Periodic Tonelli Hamiltonians   55
6.2   Weak KAM solution   57
6.3   Pseudographs, Aubry, Mañé, and Mather sets   59
6.4   The dual setting, forward solutions   60
6.5   Peierls barrier, static classes, elementary solutions   62
6.6   The forcing relation   64
6.7   The Green bundles   65

7   Perturbative weak KAM theory   66
7.1   Semi-continuity   66
7.2   Continuity of the barrier function   68
7.3   Lipschitz estimates for nearly integrable systems   70
7.4   Estimates for nearly autonomous systems   71

8   Cohomology of Aubry-Mather type   77
8.1   Aubry-Mather type and diffusion mechanisms   77
8.2   Weak KAM solutions are unstable manifolds   83
8.3   Regularity of the barrier functions   86
8.4   Bifurcation type   88

III   Proving forcing equivalence   91

9   Aubry-Mather type at the single resonance   93
9.1   The single maximum case   93
9.2   Aubry-Mather type at single resonance   94
9.3   Bifurcations in the double maxima case   96
9.4   Hyperbolic coordinates   97
9.5   Normally hyperbolic invariant cylinder 100
9.6   Localization of the Aubry and Mañé sets 102
9.7   Genericity of the single-resonance conditions 103

10 Normally hyperbolic cylinders at double resonance   106
10.1 Normal form near the hyperbolic fixed point 107
10.2 Shil'nikov's boundary value problem 108
10.3 Properties of the local maps 110
10.4 Periodic orbits for the local and global maps 114
10.5 Normally hyperbolic invariant manifolds 118
10.6 Cyclic concatenations of simple geodesics 119

11 Aubry-Mather type at the double resonance   **121**
    11.1 High-energy case 121
    11.2 Simple non-critical case 125
    11.3 Simple critical case 126
        11.3.1 Proof of Aubry-Mather type using local coordinates . . . 126
        11.3.2 Construction of the local coordinates 129

12 Forcing equivalence between kissing cylinders   **133**
    12.1 Variational problem for the slow mechanical system 133
    12.2 Variational problem for original coordinates 136
    12.3 Scaling limit of the barrier function 139
    12.4 The jump mechanism 140

IV Supplementary topics   **145**

13 Generic properties of mechanical systems on the two-torus   **147**
    13.1 Generic properties of periodic orbits 147
    13.2 Generic properties of minimal orbits 153
    13.3 Non-degeneracy at high-energy 156
    13.4 Unique hyperbolic minimizer at very high energy 158
    13.5 Generic properties at the critical energy 160

14 Derivation of the slow mechanical system   **162**
    14.1 Normal forms near maximal resonances 162
    14.2 Affine coordinate change, rescaling, and energy reduction 172
    14.3 Variational properties of the coordinate changes 177

15 Variational aspects of the slow mechanical system   **182**
    15.1 Relation between the minimal geodesics and the Aubry sets . . . 182
    15.2 Characterization of the channel and the Aubry sets 185
    15.3 The width of the channel 188
    15.4 The case $E = 0$ 190

Appendix: Notations   **195**

References   **199**

Preface

The question of Arnold diffusion has been at the center stage of Hamiltonian dynamics ever since V. I. Arnold came up with the first example in 1964, after which his name was permanently attached to this topic. The coiner of the term, B. Chirikov, originally meant it as the stochastic diffusion. However, mathematical understanding of the stochastic side has been scarce. In the current literature, Arnold diffusion mostly means topological instability.

In this book, we discuss Arnold diffusion for a smooth system with two and a half degrees of freedom. The main theorem is that Arnold diffusion occurs for a cusp-residual perturbation of an integrable system. This result was announced by J. Mather in 2003, whose proof was unfortunately unfinished before his passing. The many works on this topic in the past two decades, ours included, largely follow the blueprint laid out by him. His influence on this topic cannot be understated. With this said, we also try to incorporate the latest developments in the field and to take a different point of view. Our approach combines the geometrical and variational aspects of the theory and is heavily influenced by the weak KAM approach advanced by A. Fathi and P. Bernard. There are other works that address the same problem, most notably the works by C.-Q. Cheng, and the works by J.-P. Marco (some in collaboration with M. Gidea). We provide references and some discussions at the end of Chapter 1.

Our main goal for this book is to present a relatively self-contained full proof of this theorem. While it is not possible to give textbook-level details, we have tried to include all the important statements and proofs. Given the complex nature of the proof, we present all of its main structures in the first five chapters. It is our hope that the reader will be able to grasp the main ideas after reading these chapters.

This book has benefited from discussions with and suggestions from many people. We are grateful to the numerous discussions with J. Mather, and have benefited from the lectures he gave in the spring of 2009. The draft version has received helpful comments from A. Bounemoura, C.-Q. Cheng, and G. Forni. We deeply appreciate the efforts of the referees for their careful reading and numerous suggestions, which greatly improved the substance and presentation of the book.

Vadim Kaloshin and Ke Zhang

Acknowledgments

V. K. has been partially supported by National Science Fundation grant DMS-1402164 and enjoyed the hospitality of the ETH Institute for Theoretical Studies. He has been partially supported by Dr. Max Rössler, the Walter Haefner Foundation, and the ETH Zurich Foundation. K. Z. is supported by a National Sciences and Engineering Research Council of Canada grant, reference number REGPIN-2019-07057, and thanks the ETH Institute for Theoretical Studies for hosting him in 2017. Both authors visited the Institute of Advanced Study in the spring of 2012, and acknowledge its hospitality and a highly stimulating research atmosphere.

Part I

Introduction and the general scheme

Chapter One

Introduction

The famous question called the ergodic hypothesis, formulated by Maxwell and Boltzmann, suggests that for a typical Hamiltonian on a typical energy surface all but a set of initial conditions of zero measure, have trajectories dense in this energy surface. However, Kolmogorov-Arnold-Moser (KAM) theory showed that for an open set of (nearly integrable) Hamiltonian systems there is a set of initial conditions of positive measure with almost periodic trajectories. This disproved the ergodic hypothesis and forced reconsideration of the problem.

A quasi-ergodic hypothesis, proposed by Ehrenfest [36] and Birkhoff [17], asks if a typical Hamiltonian on a typical energy surface has a dense orbit. A definite answer to whether this statement is true or not is still far out of reach of modern dynamics. There was an attempt to prove this statement by E. Fermi [39], which failed (see [40] for a more detailed account). To simplify the problem, Arnold [4] asks:

Does there exist a real instability in many-dimensional problems of perturbation theory when the invariant tori do not divide the phase space?

For autonomous nearly integrable systems of two degrees or time-periodic systems of one and a half degrees of freedom, the KAM invariant tori divide the phase space. These invariant tori forbid large scale instability. When the degrees of freedoms are larger than two, large scale instability is indeed possible, as evidenced by the examples given by Arnold [5]. This book answers the question of the *typicality* of these instabilities in the two and a half degrees of freedom case.

1.1 STATEMENT OF THE RESULT

Let $(\theta, p) \in \mathbb{T}^2 \times B^2$ be the phase space of an integrable Hamiltonian system $H_0(p)$ with $\mathbb{T}^2$ being 2-dimensional torus $\mathbb{T}^2 = \mathbb{R}^2/\mathbb{Z}^2 \ni \theta = (\theta_1, \theta_2)$ and B^2 being the unit ball around 0 in $\mathbb{R}^2$, $p = (p_1, p_2) \in B^2$. H_0 is assumed to be strictly convex with the following uniform estimate: there exists $D > 1$ such that

$$|H_0(0)|, \quad \|\partial_p H_0(0)\| \leq D, \quad D^{-1}I \leq \partial^2_{pp} H_0(p) \leq DI, \quad \forall p \in B^2, \qquad (1.1)$$

where I is the 2×2 identity matrix.

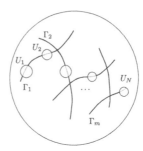

Figure 1.1: Resonant net

Consider a smooth time periodic perturbation

$$H_\epsilon(\theta, p, t) = H_0(p) + \epsilon H_1(\theta, p, t), \quad t \in \mathbb{T} = \mathbb{R}/\mathbb{T}.$$

We study Arnold diffusion for this system, namely,

topological instability in the p variable.

Arnold [5] proved existence of such orbits for an example and conjectured that they exist for a typical perturbation (see e.g. [4, 6, 7]).

Denote $\mathbb{Z}_*^3 = \mathbb{Z}^3 \setminus (0, 0, 1)\mathbb{Z}$, the integer equation $k \cdot (\omega, \omega, 1) = 0$ for $\omega \in \mathbb{R}^2$ is called a *resonance relation*. The submanifold

$$S_k = \{p \in \mathbb{R}^2 : k \cdot (\partial_p H_0, 1) = 0\}.$$

is called the resonance submanifold for k. If curves S_k and $S_{k'}$ are given by two linearly independent resonances vectors k, k', they either have no intersection or intersect at a single point in B^2. Write $k = (k^1, k^0) \in \mathbb{Z}^2 \times \mathbb{Z}$, k is called *space irreducible* if the greatest common divisor of components of k^1 is one. The union of resonances S_k with space irreducible k's form a dense subset of $\mathbb{R}^2$.

Consider a finite collection of tuples:

$$\mathcal{K} = \left\{ (k, \Gamma_k) : \quad k \in \mathbb{Z}_*^3, \quad \Gamma_k \subset S_k \cap B^2 \right\},$$

where k is *space-irreducible*, and $\Gamma_k \subset S_k$ is a closed segment. We say $\mathcal{K}$ defines a *diffusion path* if

$$\mathcal{P} = \bigcup_{(k, \Gamma_k) \in \mathcal{K}} \Gamma_k$$

is a connected set. We would like to construct diffusion orbits along the path $\mathcal{P}$ (see Figure 1.1).

Remark 1.1. The requirement of space-irreducibility is only for technical convenience, and not essential to the construction.

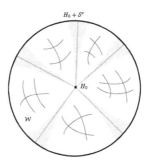

Figure 1.2: Description of cusp-generic perturbations

Theorem 1.2. *Let $5 \leq r < \infty$ and suppose $H_0 \in C^r(\mathbb{R}^2)$ satisfy condition (1.1). Let $\mathcal{P} \subset B^2$ be a diffusion path, and $U_1, \ldots, U_N$ be open sets such that $U_i \cap \mathcal{P} \neq \emptyset$, $i = 1, \ldots, N$. Then there exist:*

- *a C^r open and dense set $\mathcal{U} = \mathcal{U}(\mathcal{P}) \subset \mathcal{S}^r = \{\|H_1\|_{C^r} = 1\}$ depending on the diffusion path $\mathcal{P}$ (but not on the open sets $U_1, \ldots, U_N$),*
- *a nonnegative lower semi-continuous function $\epsilon_0 = \epsilon_0(H_1)$ with $\epsilon_0|_{\mathcal{U}} > 0$,*
- *a "cusp" set*

$$\mathcal{V} := \mathcal{V}(\mathcal{U}, \epsilon_0) := \{\epsilon H_1 : \; H_1 \in \mathcal{U}, \; 0 < \epsilon < \epsilon_0(H_1)\},$$

- *a C^r open and dense subset of $\epsilon H_1 \in \mathcal{W} \subsetneq \mathcal{V}$,*

such that for each $\epsilon H_1 \in \mathcal{W}$ there is an orbit $(\theta, p)(t)$ of H_ϵ and times $0 < T_1 < \cdots < T_N$ with the property

$$p(T_i) \in U_i, \qquad i = 1, \ldots, N.$$

Remark 1.3. The condition that ϵ_0 is lower semi-continuous implies the set $\mathcal{V}$ is open.

Remark 1.4. The set $\mathcal{W}$ may be understood as follows. The open and dense set $\mathcal{U}$ represents the set of "good directions" of perturbations, and the set of exceptional directions is nowhere dense. Around each exceptional direction we remove a cusp and call the complement $\mathcal{V}$. For this set of perturbations we establish a connected collection of invariant manifolds. Then in the complement to some exceptional perturbations $\mathcal{W}$ in $\mathcal{V}$ we show that there are diffusing orbits "shadowing" these cylinders.

This notion is due to Mather [62], who called it *cusp residual*. See Figure 1.2.

Remark 1.5. The openness of the set $\mathcal{W}$ follows immediately from the continuous dependence of the solution on the Hamiltonian. The main challenge is to prove density.

Given $(k_1, \Gamma_{k_1}) \in \mathcal{K}$ and $\lambda > 0$, we define in Section 3.1 a quantitative non-degeneracy hypothesis relative to the resonant segment Γ_{k_1}:

$$SR(k_1, \Gamma_{k_1}, \lambda) = [SR1_\lambda] - [SR3_\lambda].$$

Let us denote by $\mathcal{U}_{SR}^\lambda(k_1, \Gamma_{k_1})$ the set of H_1 that satisfies $SR(k_1, \Gamma_{k_1}, \lambda)$.

Suppose H_1 satisfies the condition $SR(k_1, \Gamma_{k_1}, \lambda)$. We define a finite subset of $\mathbb{Z}_*^3$ called the *strong additional resonances*. In Theorem 3.3 we define a large constant $K = K(k_1, \Gamma_{k_1}, \lambda)$ and call $k_2 \in \mathbb{Z}_*^3$ *strong* if:

- $(k_2, \Gamma_{k_2}) \in \mathcal{K}$ and such that $\Gamma_{k_1} \cap S_{k_2} \neq \emptyset$ and $|k_2| \leq K(k_1, \Gamma_{k_1}, \lambda)$.

We emphasize that strong additional resonances are taken from the set $\mathbb{Z}_*^3$, not just the space-irreducible ones. Denote the set of strong additional resonances $\mathcal{K}^{st}(k_1, \Gamma_{k_1}, \lambda)$. If k_2 is strong, then it defines a unique double resonance $\Gamma_{k_1,k_2} = \Gamma_{k_1} \cap S_{k_2}$.

For each double resonance Γ_{k_1,k_2}, we associate non-resonance conditions of two types:

- high energy $[DR1^h] - [DR3^h]$ (Section 4.2),
- low energy $[DR1^c] - [DR4^c]$ (Section 4.2).

For each $(k_1, \Gamma_{k_1}) \in \mathcal{K}$ and $k_2 \in \mathcal{K}^{st}(k_1, \Gamma_{k_1}, \lambda)$ consider the set of H_1 which satisfy the above conditions and denote it by $\mathcal{U}_{DR}(k_1, \Gamma_{k_1}, k_2)$.

Remark 1.6. All our non-degenerate conditions at a double resonance are stated relative to a single resonance. In particular, the condition $DR(k_1, \Gamma_{k_1}, k_2)$ may differ from the condition $DR(k_2, \Gamma_{k_2}, k_1)$. Roughly speaking, condition $DR(k_1, \Gamma_{k_1}, k_2)$ is used for diffusion along Γ_{k_1} across the double resonance Γ_{k_1,k_2}, while $DR(k_2, \Gamma_{k_2}, k_1)$ is for diffusion along Γ_{k_2} across Γ_{k_1,k_2}. When both conditions are satisfied, our diffusion mechanism also ensures the existence of an orbit that switches from Γ_{k_1} to Γ_{k_2} at the intersection Γ_{k_1,k_2}, using an equivalence relation that is sufficient for the existence of diffusion. This will be explained in Section 1.2.

The following theorem (Theorem 1.7) is immediate given that

- (Proposition 3.2) Each $\mathcal{U}_{SR}^\lambda(k_1, \Gamma_{k_1})$ is open and the union $\bigcup_{\lambda>0} \mathcal{U}_{SR}^\lambda(k_1, \Gamma_{k_1})$ is dense.
- (Proposition 4.3) The set $\mathcal{U}_{DR}(k_1, \Gamma_{k_1}, k_2)$ is open and dense.

Theorem 1.7. *For each $\lambda > 0$, the set*

$$\mathcal{U} = \mathcal{U}(\mathcal{P}) := \bigcup_{\lambda>0} \bigcap_{(k_1, \Gamma_{k_1}) \in \mathcal{K}} \left(\mathcal{U}_{SR}^\lambda(k_1, \Gamma_{k_1}) \cap \bigcap_{k_2 \in \mathcal{K}^{st}(k_1, \Gamma_{k_1}, \lambda)} \mathcal{U}_{DR}(k_1, \Gamma_{k_1}, k_2) \right)$$

is open and dense in $\mathcal{S}^r$.

As a corollary of Theorem 1.2, we obtain

Theorem 1.8 (Almost Density Theorem). *For any $\rho > 0$ there are*

- *an open dense set $\mathcal{U} = \mathcal{U}(\rho) \subset \mathcal{S}^r$,*
- *a non-negative lower semi-continuous function $\epsilon_0 : \mathcal{S}^r \to \mathbb{R}_+$ with $\epsilon_0|_\mathcal{U} > 0$,*
- *a cusp set $\mathcal{V} := \mathcal{V}(\mathcal{U}, \epsilon_0) := \{\epsilon H_1 : H_1 \in \mathcal{U}, \ 0 < \epsilon < \epsilon_0(H_1)\}$,*
- *an open dense subset $\mathcal{W} \subsetneq \mathcal{V}$,*

such that for any $\epsilon H_1 \in \mathcal{W}$ there is a ρ-dense orbit on $\mathbb{T}^2 \times B^2 \times \mathbb{T}$.

Proof. Suppose $p_* \in B^2$ and let $\omega = \nabla H_0(p_*)$. We say ω is ρ-irrational if there exists $T > 0$ such that $\{t(\omega, 1) : t \in [-T, T]\} \subset \mathbb{T}^3$ is ρ-dense, and let $T(\omega)$ be the smallest such T. Using the fact that $\dot{p} = O(\epsilon)$, $\dot{\theta} = \nabla H_0(p) + O(\epsilon)$, there is $\epsilon_0 > 0$ depending on ρ, $T(\omega)$ and $\|H_0\|_{C^2}$ such that if $0 < \epsilon < \epsilon_0$,

$$B_{2\rho}\left(\bigcup_{t \in \mathbb{R}} \phi^t_{H_\epsilon}(\theta_0, p_0, t_0)\right) \supset \mathbb{T}^2 \times B_\rho(p_*) \times \mathbb{T}, \tag{1.2}$$

$$\text{for all } p_0 \in B_\rho(p_*), (\theta_0, t_0) \in \mathbb{T}^2 \times \mathbb{T}.$$

A resonance $k \in \mathbb{Z}^3_*$ is called ρ-irrational if the sub-torus $\{x \in \mathbb{T}^3 : x \cdot k = 0\}$ is ρ-dense. Dirichlet's Theorem implies that there are at most finitely many $k \in \mathbb{Z}^3_*$ which is not ρ-irrational (called ρ-rational). We claim that if k is ρ-irrational, then the ρ-irrational frequencies form a dense subset of $S_k = \{\omega \in \mathbb{R}^2 : (\omega, 1) \cdot k = 0\}$. Indeed, if $\omega \in S_k$ is ρ-rational, then $\mathbb{T}_\omega = \overline{\{t(\omega, 1) : t \in \mathbb{R}\}}$, which is not ρ-dense and is a sub-torus of dimension 1 or 2. If $\dim \mathbb{T}_\omega = 2$, we must have $\mathbb{T}_\omega = \mathbb{T}_k$, which is impossible, since $\mathbb{T}_k$ is ρ-dense. Therefore $\dim \mathbb{T}_\omega = 1$ and $(\omega, 1)$ is doubly resonant. The set of such ω is at most countable.

Since there are infinitely many space-irreducible resonances, there is a diffusion path $\mathcal{P}$ consisting only of $\rho/2$-irrational resonances. Moreover, we may choose the path $\mathcal{P}$ to be $\rho/2$-dense in B^2, since space-irreducible resonances are dense.

We now apply Theorem 1.2 to the path $\mathcal{P}$, and pick p_i, $i = 1, \ldots, N \in \mathcal{P}$ such that $(\nabla H_0(p_i), 1)$ is $(\rho/2)$-irrational, and such that $\bigcup_{i=1}^n B_{\rho/2}(p_i) \supset B^2$. According to our theorem, there is an orbit whose p component visits every $B_\rho(p_i)$. Then the orbit must be ρ-dense in view of (1.2). $\qquad \square$

1.2 SCHEME OF DIFFUSION

A Hamiltonian $H : \mathbb{T}^2 \times \mathbb{R}^2 \times \mathbb{T} \to \mathbb{R}$ is called *Tonelli* if it satisfies:

- (Strictly convex) For each $p \in \mathbb{R}^2$, $\partial^2_{pp} H(\theta, p, t)$ is strictly positive definite.
- (Super linear) $\lim_{\|p\| \to \infty} H(\theta, p, t)/\|p\| = \infty$.
- (Complete) The Hamiltonian flow generated by H is complete.

If H_0 satisfies condition (1.1), then for all $\epsilon < D^{-1}$, $H_\epsilon = H_0 + \epsilon H_1$ are Tonelli. Indeed, convexity and super-linearity are immediate, while the flow of H_ϵ is complete, since the Hamiltonian vector field is uniformly Lipschitz.

A Tonelli Hamiltonian can always be associated with a Lagrangian

$$L(\theta, v, t) = \sup_{p \in \mathbb{R}^2} \{p \cdot v - H(\theta, p, t)\},$$

and one can study the dynamics in the Lagrangian setting. Using Lagrangian dynamics, Mather developed a theory ([37], [61]) for special invariant sets of Tonelli Hamiltonians. He showed that for each cohomology class $c \in \mathbb{R}^2 \simeq H^1(\mathbb{T}^2, \mathbb{R})$, the Hamiltonian H_ϵ admits families of invariant sets of the Hamiltonian flow on $\mathbb{T}^2 \times \mathbb{R}^2 \times \mathbb{T}$, called the Mather, Aubry, and Mañé sets. In our perturbative setting, one has the relation

$$\widetilde{\mathcal{M}}_{H_\epsilon}(c) \subset \tilde{\mathcal{A}}_{H_\epsilon}(c) \subset \tilde{\mathcal{N}}_{H_\epsilon}(c) \subset \mathbb{T}^2 \times B_{C\sqrt{\epsilon}}(c) \times \mathbb{T},$$

where C is a constant depending only on D (see Corollary 7.7). Their projections to the (θ, t) components are denoted

$$\mathcal{M}_{H_\epsilon}(c), \quad \mathcal{A}_{H_\epsilon}(c), \quad \mathcal{N}_{H_\epsilon}(c).$$

We use $\widetilde{\mathcal{M}}^0_{H_\epsilon}(c)$, $\tilde{\mathcal{A}}^0_{H_\epsilon}(c)$, and $\tilde{\mathcal{N}}^0_{H_\epsilon}(c)$ to denote their intersections with the time-0 section $\{t = 0\}$, which are invariant under the time-1 map. Throughout the book, we may switch between the two equivalent settings: consider either continuous invariant sets of the flow, or discrete invariant sets under the time-1 map.

Our main strategy is to pick a subset $\Gamma_* \subset \mathbb{R}^2$ of cohomologies very close to the diffusion path $\mathcal{P}$, then find an orbit that shadows a sequence of Aubry sets $\tilde{\mathcal{A}}_{H_\epsilon}(c_i)$, $i = 1, \ldots, N$, $c_i \in \Gamma_*$. We show that these Aubry sets admit non-degenerate *heteroclinic connections* between them, called a *transition chain* by Arnold ([5]). The existence of these connections from one of the four *diffusion mechanisms*. We give a general introduction to these mechanisms in the text that follows, and refer to Section 2.1 for precise definitions.

Mather mechanism

In the case that H_ϵ is defined on $\mathbb{T} \times \mathbb{R} \times \mathbb{T}$, the time-1 map of H_ϵ is a monotone twist map on the infinite cylinder $\mathbb{T} \times \mathbb{R}$. An essential invariant curve is an invariant curve that is homotopic to $\mathbb{T} \times \{0\}$. It is known since Birkhoff that a region free of essential invariant curves is unstable, namely there exist orbits that drift from one boundary of the region to another. Mather [61] gave a conceptual description of this phenomenon and generalized it to higher dimensions.

We say that the pair (H_ϵ, c) satisfies the *Mather mechanism* if

$$\mathcal{N}^0_{H_\epsilon}(c) = \pi_\theta \tilde{\mathcal{N}}^0_{H_\epsilon}(c) \subset \mathbb{T}^2$$

is contractible, where $\pi_\theta : \mathbb{T}^2 \times \mathbb{R}^2$ is the projection to the first component. (In the twist map case this means $\widetilde{\mathcal{N}}^0(c)$ is not an essential invariant curve.) Mather proved that ([61]) in this case, $\widetilde{\mathcal{A}}^0_{H_\epsilon}(c)$ admits a heteroclinic connecting orbit to $\widetilde{\mathcal{A}}^0_{H_\epsilon}(c')$ if c, c' are close.

Arnold mechanism

In Arnold's original paper [5], Arnold showed the existence of a family of invariant tori, whose own stable and unstable manifolds intersect transversally. In our setting, the tori are the Aubry sets $\widetilde{\mathcal{A}}_{H_\epsilon}(c)$ contained in a 3-dimensional *normally hyperbolic invariant cylinder* (which means homotopic to $\mathbb{T}^2 \times \mathbb{R}$). Then $\widetilde{\mathcal{A}}^0_{H_\epsilon}(c)$ is contained in a 2-dimensional cylinder. To consider homoclinic connections, we lift the system to a double covering map, then a homoclinic connection of $\widetilde{\mathcal{A}}_{H_\epsilon}(c)$ becomes a *heteroclinic* connection between the two copies.

We say that the pair (H_ϵ, c) satisfy the *Arnold mechanism* if $\widetilde{\mathcal{A}}^0_{H_\epsilon}(c)$ is an invariant curve and there exists a symplectic double covering map $\Xi : \mathbb{T}^2 \times \mathbb{R}^2 \to \mathbb{T}^2 \times \mathbb{R}^2$, such that the set

$$\widetilde{\mathcal{N}}^0_{H_\epsilon \circ \Xi}(\Xi^* c) \setminus \widetilde{\mathcal{A}}^0_{H_\epsilon \circ \Xi}(\Xi^* c)$$

is totally disconnected. If $\widetilde{\mathcal{A}}^0_{H_\epsilon}(c)$ is a smooth invariant curve with transversal intersection of stable and unstable manifolds, then the above set is discrete.

This mechanism is inspired by Arnold's pioneering paper [5], hence its name. In our setting, Cheng-Yan [26, 27] and Bernard [11] showed that if (H_ϵ, c) satisfy the Arnold mechanism, $\widetilde{\mathcal{A}}^0_{H_\epsilon}(c)$ admits a heteroclinic connecting orbit to $\widetilde{\mathcal{A}}^0_{H_\epsilon}(c')$ if c, c' are close.

Bifurcation mechanism

This is technically similar to the Arnold mechanism, but happens when the Aubry set $\widetilde{\mathcal{A}}_{H_\epsilon}(c)$ is contained in two disjoint normally hyperbolic invariant cylinders.

We say that the pair (H_ϵ, c) satisfies the *bifurcation mechanism* if the set

$$\widetilde{\mathcal{N}}^0_{H_\epsilon}(c) \setminus \widetilde{\mathcal{A}}^0_{H_\epsilon}(c)$$

is totally disconnected. In this case, the same conclusion as in the Arnold mechanism holds.

Normally hyperbolic invariant cylinders and Aubry-Mather types

The three mechanisms given do not apply to all cases, even after a generic perturbation. The main observation is that they do apply to cohomologies that satisfy:

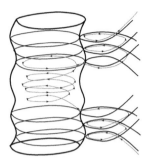

Figure 1.3: Diffusion along a cylinder.

1. (2D NHIC) The Aubry set $\widetilde{\mathcal{A}}^0_{H_\epsilon}(c)$ is contained in a 2-dimensional normally hyperbolic invariant cylinder.
2. (1D Graph Theorem) The Aubry set $\widetilde{\mathcal{A}}^0_{H_\epsilon}(c)$ is contained in a Lipschitz graph over the circle $\mathbb{T}$.

In this case, we say that the pair (H_ϵ, c) is of the *Aubry-Mather type*.

Under these assumptions, the Aubry set resembles the *Aubry-Mather sets* for twist maps, and in particular, generically we have the following dichotomy: either $\pi\widetilde{\mathcal{N}}^0_{H_\epsilon}(c)$ is contractible or $\widetilde{\mathcal{A}}^0_{H_\epsilon}(c)$ is a Lipschitz invariant curve. In the latter case, we can show that the Arnold mechanism applies after an additional perturbation. Since either the Mather or Arnold mechanism applies, we conclude that $\widetilde{\mathcal{A}}(c)$ is connected to $\widetilde{\mathcal{A}}(c')$ for c, c' close. Moreover, this argument can be continued if c' is also of Aubry-Mather type. Dynamically, the orbit is either diffusing along the heteroclinic orbits of invariant curves or diffusing in a Birkhoff region of instability within the cylinder. See Figure 1.3.

While a cohomology of Aubry-Mather type is robust, namely it can be extended along a continuous curve, in a one-parameter family one may encounter a bifurcation where the Aubry set jumps from one cylinder to another one. At the bifurcation, the Aubry set is contained in both cylinders. We say that the pair (H_ϵ, c) is of *bifurcation Aubry-Mather type* if the Aubry set is possibly contained in two cylinders.

For technical reasons, we have to involve a different bifurcation type, called *asymmetric bifurcation type*. This is very similar to the bifurcation Aubry-Mather type, the main difference is, on one side of the bifurcation, the Aubry set is an Aubry-Mather type set contained in an invariant cylinder, while on the other side we have a hyperbolic periodic orbit. This happens when we cross double resonance.

Forcing relation and jump mechanism

The rigorous formulation of the three diffusion mechanisms will be given using the concept of *forcing equivalence* defined by Bernard in [11], which is a generalization of an equivalence relation defined by Mather; see [61]. If c, c' are forcing equivalent (denoted $c \dashv\vdash c'$), then there is a heteroclinic orbit connecting the associated Aubry sets. Moreover, there exists an orbit shadowing an arbitrary sequence of cohomologies, as long as they are all equivalent. See Section 2.1 for more details.

The main theorem reduces to Theorem 2.2, which proves forcing equivalence of a net of cohomologies, called Γ_*. The set Γ_* consists of finitely many smooth curves. On each of the smooth curves, we prove the cohomologies are of Aubry-Mather type, and therefore one of the three mechanisms applies, after a generic perturbation.

We prove forcing equivalence of different connected components directly, using the definition of the forcing relation. We call this the **jump mechanism**.

1.3 THREE REGIMES OF DIFFUSION

Our plan is to choose a net Γ_* of cohomology classes, and prove their forcing equivalence, by first proving they are of Aubry-Mather type or bifurcation Aubry-Mather type. This is done in three distinct regimes.

Single resonance

Let $(k_1, \Gamma_{k_1}) \in \mathcal{K}$ be one of the single-resonance component, and $\mathcal{K}^{\mathrm{st}}(k_1, \Gamma_{k_1}, \lambda)$ be the collection of the strong additional resonances. Then for p in a $O(\sqrt{\varepsilon})$-neighborhood of the set

$$
\Gamma_{k_1}^{SR}(M, \lambda) := \Gamma_{k_1} \setminus \left(\bigcup_{k_2 \in \mathcal{K}^{\mathrm{st}}(k_1, \Gamma_{k_1}, \lambda)} B_{M\sqrt{\epsilon}}(\Gamma_{k_1, k_2}) \right),
$$

where M is a large parameter, the system admits the normal form

$$
N_\epsilon^{SR} = H_0 + \epsilon Z(\theta^s, p) + O(\epsilon\delta), \quad (\theta^s, \theta^f, t) \in \mathbb{T}^3,
$$

where how small δ is depends on how many double resonances we exclude. Under the non-degeneracy conditions $SR(k_1, \Gamma_{k_1}, \lambda)$, the above system admits 3-dimensional (for the flow) normally hyperbolic invariant cylinders, and one can prove each $c \in \Gamma_{k_1}^{SR}(M, \lambda)$ is of AM or of bifurcation AM type.

Double resonance, high energy

Let $p_0 = \Gamma_{k_1, k_2}$ be a strong double resonance. On the set

$$
B_{M\sqrt{\epsilon}}(p_0), \quad p_0 = \Gamma_{k_1} \cap \Gamma_{k_2}
$$

we perform a normal form transformation and then rescale the p variable via $I = (p - p_0)/\sqrt{\epsilon}$. One can show that the system is conjugate to

$$\frac{1}{\beta} \left(K(I) - U(\varphi) + O(\sqrt{\epsilon}) \right),$$

where $K : \mathbb{R}^2 \to \mathbb{R}$ is a positive quadratic form and $U : \mathbb{T}^2 \to \mathbb{R}$, and $\beta > 0$ is a constant depending only on k_1, k_2. The system $H^s = K(I) - U(\varphi)$ is a two degrees of freedom mechanical system, and is called the *slow mechanical system*. In the text that follows we use the shifted energy $E := H^s(\varphi, I) - \min H^s$ as a parameter. By the Maupertuis principle, for each $E > 0$, the Hamiltonian flow in the energy surface $H^s - \min H^s = E$ is conjugate to a geodesic flow, whose metric is called the Jacobi metric g_E.

When the shifted energy E is not too close to 0, we are in the *high-energy regime*. By imposing the conditions $[DR1^h] - [DR3^h]$, one shows the existence of 2-dimensional normally hyperbolic invariant cylinders associated to the shortest loops for the associated Jacobi metric. This cylinder persists under perturbation, and one can show that the associated cohomologies are of Aubry-Mather type.

Double resonance, low energy

As the energy decreases, the cylinder constructed in the high-energy regime may not persist. Under the non-degeneracy conditions $DR(k_1, k_2)$, we distinguish two separate cases:

1. Simple cylinder: In this case the cylinder extends across zero shifted energy to negative shifted energy. In this case one can still show the associated cohomologies are of Aubry-Mather type.
2. Non-simple cylinder: In this case the cylinder may be destroyed before the shifted energy becomes zero. However, we show the existence of two simple cylinders near the non-simple one, and one can "jump" from one cylinder to another one.
   This is the only case where the **jump mechanism** is used.

1.4   THE OUTLINE OF THE PROOF

We structured this book into four parts:

Part I (Chapters 1 to 5)

We formulate all the major definitions and theorems used in the main proof, and prove our main theorem (Theorem 1.2) assuming results proven in the later chapters. We hope the reader will be able to get a good idea of the whole proof by just reading Part I.

Chapter 2 describes forcing relations, different diffusion mechanisms, and Aubry-Mather types. It also formulates Theorem 2.2 and shows that it implies our main theorem.

Chapter 3 first describes the single resonance non-degeneracy conditions and normal forms. It then formulates Theorem 3.3, which covers the forcing equivalence in the single resonance regime.

The double resonance is split into two chapters. Chapter 4 describes the geometrical structure for the system at double resonance. After the normal form near a double resonance is desribed, the system is reduced to the slow mechanical system with perturbation. We formulate the non-degeneracy conditions and theorems about their genericity. The normally hyperbolic invariant cylinders are described, and the proof using local and global maps is sketched.

Chapter 5 describes the choice of cohomology and Aubry-Mather type at the double resonance. First, cohomology classes are chosen for the (unperturbed) slow mechanical system. We prove Aubry-Mather type for the perturbed slow mechanical system and revert to the original coordinates. As the system has been perturbed, we need to modify the choice of cohomology classes to connect the single and double resonances. At the end of this chapter, Theorem 2.2 is proved, proving our main theorem assuming statements proved in the rest of this book.

Part II (Chapters 6 to 8)

We introduce most of the variational theory used in this book. Chapter 6 is an overview of standard weak KAM theory, Chapter 7 concerns perturbation aspects of the weak KAM theory, and Chapter 8 defines cohomology of Aubry-Mather type and explains why it implies one of the diffusion mechanisms, after a generic perturbation.

Part III (Chapters 9 to 12)

We prove that the Aubry-Mather type holds in various regimes, and also describes the *jump mechanism*. Chapter 9 proves Aubry-Mather type in the single-resonance regime, Chapter 10 proves the geometric picture of double-resonance described in Chapter 4, and Chapter 11 proves Aubry-Mather type for the double resonance regime. The jump mechanism is formulated and proved in Chapter 12.

Part IV (Chapters 13 to 15)

These chapters include topics that more or less stand alone. Chapter 13 proves genericity of the double resonance conditions. The flavor is similar to that of Kupka-Smale theorem. Chapter 14 proves various normal form results, and formulates the coordinate changes that are used to derive the slow system at the double resonance. Chapter 15 describes the variational property of the slow mechanical system. Much of this topic is covered in Mather's works [64, 65], but we give a self-contained presentation.

1.5   DISCUSSION

Relation with Mather's approach

Theorem 1.2 was announced by Mather in [62], where he proposed a plan to prove it. Some parts of the proof are provided in [63]. Our work realizes Mather's general plan using weak KAM theory and Hamiltonian point of view. In the following we summarize the new techniques and tools introduced.

- We utilize Bernard's forcing relation to simplify the construction of a diffusion orbit. This allows a more Hamiltonian treatment of the variational concepts and allows us to reduce the main theorem to local forcing equivalence of cohomology classes.
- We use Hamiltonian normal forms to construct a collection of normally hyperbolic invariant cylinders along the chosen diffusion path. We obtain precise control of the normal forms (via an anisotropic C^2 norm) at both single and double resonances. Mather's method uses mostly the Lagrangian point of view.
- We introduce the concept of *Aubry-Mather type*, which generalizes the work done in Bernard-Kaloshin-Zhang [13] to a more abstract setting, applicable to both single and double resonances. Heuristically, this means the Aubry set has the same topological structures as Aubry-Mather sets in twist maps, in particular, the set has a proper ordering. Our approach can be seen as a generalization of the variational technique for a priori unstable systems from [11, 13, 27].
- One important obstacle is the problem of regularity of barrier functions (see Section 8.3), which outside of the realm of twist maps is difficult to overcome. Our definition of Aubry-Mather type allows proving this statement in a general setting. It is our understanding that Mather [58] handles this problem without proving the existence of invariant cylinders.
- In a double resonance we also construct normally hyperbolic invariant cylinders. This leads to a fairly simple and explicit structure of minimal orbits near a double resonance. In particular, in order to switch from one resonance to another we need *only one jump* (see Section 5.5 for the formulation of the statement).
- It is our understanding that Mather's approach [58] requires an implicitly defined number of jumps. His approach resembles his proof of existence of diffusing orbits for twist maps inside a Birkhoff region of instability [59].

Other results on a priori stable systems

A different approach to prove a theorem of similar flavor using variational methods is given in the series of works [24, 23, 25]. In [54] (see also [44, 55]), the authors develop a geometric method to prove Arnold diffusion.

An important aspect for studying a nearly integrable system is a partial averaging theory based on arithmetic properties of the frequency. We develop a version of this theory in [52].

Generic instability of resonant totally elliptic points

In [50] stability of resonant totally elliptic fixed points of symplectic maps in dimension 4 is studied. It is shown that generically a convex, resonant, totally elliptic point of a symplectic map is Lyapunov unstable.

Non-convex Hamiltonians

In the case the Hamiltonian H_0 is non-convex or non-strictly convex for all $p \in B^2$, for example, $H_0(p) = p_1^2 + p_2^3$, the problem of global Arnold diffusion is wide open. Some results for the Hamiltonian $H_0(p) = p_1^2 - p_2^2$ are in [16, 20].

To apply a variational approach one faces another deep open problem of extending Mather theory and weak KAM theory beyond convex Hamiltonians or developing a new technique to construct diffusing orbits.

Other diffusion mechanisms

Here we would like to give a short review of other diffusion mechanisms. In the case $n = 2$ Arnold proposed the following example:

$$H(q, p, \phi, I, t) = \frac{I^2}{2} + \frac{p^2}{2} + \epsilon(1 - \cos q)(1 + \mu(\sin \phi + \sin t)).$$

This example is a perturbation of the product of a 1-dimensional pendulum and a 1-dimensional rotator. The main feature of this example is that it has a 3-dimensional normally hyperbolic invariant cylinder. There is a rich literature on the Arnold example and we do not intend to give n extensive list of references; we mention [2, 9, 14, 15, 79] and references therein. This example gave rise to a family of examples of systems of $n + 1/2$ degrees of freedom of the form

$$H_\epsilon(q, p, \phi, I, t) = H_0(I) + K_0(p, q) + \epsilon H_1(q, p, \phi, I, t),$$

where $(q, p) \in \mathbb{T}^{n-1} \times \mathbb{R}^{n-1}, I \in \mathbb{R}, \phi, t \in T$. Moreover, the Hamiltonian $K_0(p, q)$ has a saddle fixed point at the origin and $K_0(0, q)$ attains its strict maximum at $q = 0$. For small ϵ a 3-dimensional NHIC $\mathcal{C}$ persists. Several geometric mechanisms of diffusion have evolved.

In [31, 32, 33, 43] the authors carefully analyze two types of dynamics induced on the cylinder $\mathcal{C}$. These two dynamics are given by so-called inner and outer maps. In [35, 34], these techniques are applied to a general perturbation of Arnold's example.

In [30, 75, 76, 77] a return (separatrix) map along invariant manifolds of $\mathcal{C}$ is constructed. A detailed analysis of this separatrix map gives diffusing orbits.

In [22, 45, 51] for an open set of perturbations of Arnold's example, one constructs a probability measure μ in the phase space such that the pushforward of μ projected onto the I component in the proper time scale weakly converges to the stochastic diffusion process. This, in particular, implies the existence of diffusing orbits.

In [42], the authors treats the a priori chaotic setting, but prove diffusion in the real analytic category, which is much more difficult. A different mechanism related to the slow–fast system is given by the same authors in [41].

For a priori unstable systems the works [11, 26, 27] are inspired and influenced by Mather's approach [59, 60, 61] and build diffusing orbits variationally. Recentlyan a priori unstable structure was established for the restricted planar three-body problem [38]. It turns out that for this problem there are no large gaps.

A multidimensional diffusion mechanism of different nature, but also based on existence and persistence of a 3-dimensional NHIC $\mathcal{C}$ is proposed in [20].

There are many studies of Arnold diffusion using computer numerics: see for example [46] and the references therein.

Chapter Two

Forcing relation

2.1 SUFFICIENT CONDITION FOR ARNOLD DIFFUSION

We will utilize the concept of forcing equivalence, denoted $c \dashv\vdash c'$. The actual definition will not be important for the current discussions, instead, we state its main application to Arnold diffusion.

Proposition 2.1 (Proposition 0.10 of [11]). *Let $\{c_i\}_{i=1}^N$ be a sequence of cohomology classes which are forcing equivalent with respect to H. For each i, let U_i be neighborhoods of the discrete Mather sets $\widetilde{\mathcal{M}}_H^0(c_i)$. Then there is a trajectory of the Hamiltonian flow visiting all the sets U_i in the prescribed order.*

Let $\sigma > 0$ and $\mathcal{V}_\sigma(H)$ denote the σ neighborhood of H in the space $C^r(\mathbb{T}^2 \times B^2 \times \mathbb{T})$ with respect to the natural C^r topology. The following statement is a "local" version of our main theorem, where we state that given $H_1 \in \mathcal{U}$, we can

- Choose ϵ_0 to be locally constant on a neighborhood of H_1
- Prove forcing equivalence on a residual subset of a neighborhood of H_ϵ.

Theorem 2.2. *Let $\mathcal{P}$ be a diffusion path and $U_1, \ldots, U_N$ be open sets intersecting $\mathcal{P}$. Then there is an open and dense subset $\mathcal{U}(\mathcal{P}) \subset \mathcal{S}^r$, and for each $H_1 \in \mathcal{U}(\mathcal{P})$, there are $\delta = \delta(H_0, H_1) > 0$, $\epsilon_1 = \epsilon_1(H_0, H_1) > 0$ such that for each*

$$H_1' \in \mathcal{V}_\delta(H_1), \quad 0 < \epsilon < \epsilon_1,$$

there is a subset $\Gamma_*(\epsilon, H_0, H_1') \subset \mathbb{R}^2$ satisfying

$$\Gamma_* = \Gamma_*(\epsilon, H_0, H_1') \cap U_i \neq \emptyset, \quad i = 1, \ldots, N,$$

with the property that there is $\sigma = \sigma(\epsilon, H_0, H_1') > 0$ and a residual subset $\mathcal{R}_\sigma(H_0 + \epsilon H_1') \subset \mathcal{V}_\sigma(H_0 + \epsilon H_1')$, such that for each $H' \in \mathcal{R}$, with respect to the Hamiltonian H', all the $c \in \Gamma_*(\epsilon, H_0, H_1')$ are forcing equivalent to each other.

Proposition 2.1 and Theorem 2.2 imply our main theorem.

Proof of Theorem 1.2. First of all, let us define the lower semi-continuous function ϵ_0. For each $H_1 \in \mathcal{U}$, define

$$\epsilon_2^{H_1}(\cdot) = \epsilon_1(H_0, H_1)\mathbf{1}_{\mathcal{V}_\delta(H_1)}(\cdot),$$

where $\mathbf{1}_\mathcal{V}$ denote the indicator function of $\mathcal{V}$. An indicator function of an open set is lower semi-continuous by definition. For $H_1 \in \mathcal{S}^r \setminus \mathcal{U}$, let $\epsilon_2^{H_1} \equiv 0$. We then define

$$\epsilon_0(\cdot) = \sup_{H_1 \in \mathcal{S}^r} \epsilon_2^{H_1}(\cdot),$$

which is lower semi-continuous, being an (uncountable) supremum of lower semi-continuous functions. Note that ϵ_0 is positive on each $H_1 \in \mathcal{U}$ since $\epsilon_2^{H_1}(H_1) = \epsilon_1(H_0, H_1) > 0$.

Consider $H_1 \in \mathcal{U}$ and $0 < \epsilon < \epsilon_0(H_1)$ as defined earlier. Let $\Gamma_*(\epsilon_0, H_0, H_1)$ be as in Theorem 2.2. Let $c_i \in U_i \cap \Gamma_*(\epsilon, H_0, H_1)$. For any $\|H - H_0\|_{C^r} \le \epsilon$, by Corollary 7.7, there is $C > 0$ depending only on D such that $\widetilde{\mathcal{M}}_H^0(c) \subset \mathbb{T}^2 \times B_{C\sqrt{\epsilon}}(c)$. As a result, reducing ϵ_0 if necessary (note that minimum of an lower semi-continuous function and a constant is still lower semi-continuous), we have $\widetilde{\mathcal{M}}_H^0(c_i) \subset \mathbb{T}^2 \times U_i$. Since c_i are all forcing equivalent by Theorem 2.2, Proposition 2.1 implies the existence of an orbit visiting each neighborhood $\mathbb{T}^2 \times U_i$.

Since the above discussion applies to all $H \in \mathcal{R}_\sigma(H_0 + \epsilon H_1)$, where $H_1 \in \mathcal{U}$ and $0 < \epsilon < \epsilon_0(H_1)$, we conclude that for a dense subset of $\mathcal{V}(\mathcal{U}, \epsilon_0)$ (as defined in Theorem 1.2), there is an orbit visiting each $\mathbb{T}^2 \times U_i$. Since this property is open due to the smoothness of the flow, it holds on an open and dense subset $\mathcal{W}$ of $\mathcal{V}$.   □

The set $\Gamma_*(\epsilon, H_0, H_1)$ will be chosen to be the union of finitely many smooth curves, and will coincide with $\mathcal{P}$ except on finitely many neighborhoods of size $O(\sqrt{\epsilon})$ of strong double resonances.

2.2   DIFFUSION MECHANISMS VIA FORCING EQUIVALENCE

We reformulate the diffusion mechanisms introduced in Section 1.2 using forcing equivalence. We start with the Mather mechanism.

Proposition 2.3 (Theorem 0.11 of [11]). *Suppose*

$$\mathcal{N}_H^0(c) \text{ is contractible} \tag{Ma}$$

as a subset of $\mathbb{T}^2$, *then there is* $\sigma > 0$ *such that* c *is forcing equivalent to all* $c' \in B_\sigma(c)$.

To define the Arnold mechanism, we consider a finite covering of our space. Let

$$\xi : \mathbb{T}^n \to \mathbb{T}^n$$

be a linear double covering map, for example: $(\theta_1, \theta_2) \mapsto (2\theta_1, \theta_2)$. Then ξ lifts to a symplectic map

$$\Xi : \mathbb{T}^n \times \mathbb{R}^n \to \mathbb{T}^n \times \mathbb{R}^n, \quad \Xi(\theta, p) = (\xi\theta, \xi^* p),$$

where $\xi^*(p)$ is defined by the relation $\xi^*(p) \cdot v = p \cdot d\xi(v)$ for all $v \in \mathbb{R}^2$. For example, if $n = 2$ and $\xi(\theta_1, \theta_2) = (2\theta_1, \theta_2)$ we have $\xi^*(p_1, p_2) = (p_1/2, p_2)$. This allows us to consider the lifted Hamiltonian $H \circ \Xi$.

Lemma 2.4. *(Section 7 in [11]) We have*

$$\widetilde{\mathcal{A}}^0_{H \circ \Xi}(\xi^* c) = \Xi^{-1} \widetilde{\mathcal{A}}^0_H(c), \quad \widetilde{\mathcal{N}}^0_{H \circ \Xi}(\xi^* c) \supset \Xi^{-1} \widetilde{\mathcal{N}}^0_H(c).$$

Moreover, $\xi^* c \vdash \xi^* c'$ relative to $H \circ \Xi$ implies $c \vdash c'$ relative to H.

The Aubry set can be decomposed into disjoint invariant sets called *static classes*, which gives important insight into the structure of the Aubry set. In particular, when there is only one static class, then $\widetilde{\mathcal{A}}^0_H(c) = \widetilde{\mathcal{N}}^0_H(c)$. In the case $\widetilde{\mathcal{A}}^0_H(c) \neq \widetilde{\mathcal{N}}^0_H(c)$, the difference $\widetilde{\mathcal{N}}^0_H(c) \backslash \widetilde{\mathcal{A}}^0_H(c)$ consists of heteroclinic orbits from one static class to another [11]. Using Lemma 2.4, when $\widetilde{\mathcal{A}}^0_H(c) = \widetilde{\mathcal{N}}^0_H(c)$, it may happen that $\widetilde{\mathcal{A}}^0_{H \circ \Xi}(\xi^* c) \subsetneq \widetilde{\mathcal{N}}^0_{H \circ \Xi}(\xi^* c)$, and the difference provides additional heteroclinic orbits to the Aubry set that are not contained in the Mañé set before the lifting. This can be exploited to create diffusion orbits.

Proposition 2.5 (Proposition 7.3 and Theorem 9.2 of [11]). *Suppose, either:*

$$\widetilde{\mathcal{A}}^0_H(c) \text{ has two static classes, and } \widetilde{\mathcal{N}}^0_H(c) \setminus \widetilde{\mathcal{A}}^0_H(c) \text{ is totally disconnected,} \tag{Bif}$$

or

$$\widetilde{\mathcal{A}}^0_H(c) \text{ has one static class, and } \widetilde{\mathcal{N}}^0_{H \circ \Xi}(\xi^* c) \setminus \widetilde{\mathcal{A}}^0_{H \circ \Xi}(\xi^* c) \text{ is totally disconnected.} \tag{Ar}$$

Then there is $\sigma > 0$ such that c is forcing equivalent to all $c' \in B_\sigma(c)$.

As implied by the labeling, the first item is called the bifurcation mechanism, and the second the Arnold mechanism. We obtain the following immediate corollary:

Corollary 2.6 (Mather-Arnold mechanism). *Suppose $\Gamma \subset B^2$ is a connected set, and for each $c \in \Gamma$ one of the diffusion mechanisms (Ma), (Bif), or (Ar) holds. Then all $c \in \Gamma$ are forcing equivalent.*

Recall that $\Gamma_*(\epsilon, H_0, H_1)$ can be chosen as a union of finitely many smooth curves.

- We will later show that for $c \in \Gamma_*(\epsilon, H_0, H_1)$ in Theorem 2.2, one of the two applies: Proposition 2.3 or Proposition 2.5.
  As a result, each connected component of $\Gamma_*(\epsilon, H_0, H_1)$ consists of equivalent

c's.

- We prove the forcing equivalence between different connected components using directly the definition of forcing relation. We call this *the "jump" mechanism.*

2.3   INVARIANCE UNDER THE SYMPLECTIC COORDINATE CHANGES

A diffeomorphism $\Psi = \Psi(\theta, p) : \mathbb{T}^n \times \mathbb{R}^n \to \mathbb{T}^n \times \mathbb{R}^n$ is called exact symplectic if $\Psi^*\lambda - \lambda$ is an exact one-form, where $\lambda = \sum_{i=1}^n p_i d\theta_i$ is the canonical form. We say $\Phi : \mathbb{T}^n \times \mathbb{R}^n \times \mathbb{T} \to \mathbb{T}^n \times \mathbb{R}^n \times \mathbb{T}$ is *exact symplectic* if there exists a smooth map $\Phi_1 : \mathbb{T}^n \times \mathbb{R}^n \times \mathbb{T} \to \mathbb{T}^n \times \mathbb{R}^n$ such that

$$\Phi(\theta, p, t) = (\Phi_1(\theta, p, t), t),$$

and there is $\widetilde{E} = \widetilde{E}(\theta, p, t)$, such that

$$\Psi(\theta, p, t, E) = (\Phi(\theta, p, t), E + \widetilde{E}(\theta, p, t)) \tag{2.1}$$

(called the autonomous extension of Φ) is exact symplectic. The new term $\widetilde{E}(\theta, p, t)$ is defined up to adding a function $f'(t)$, where $f(t)$ is periodic in t. Let us assume $\widetilde{E}(0, 0, t) \equiv 0$; therefore the choice of $\widetilde{E}$ is unique.

Let $H = H(\theta, p, t)$, and Φ is exact symplectic with extension Ψ. Then for $G(\theta, p, t, E) = H(\theta, p, t) + E$, we define

$$\Phi^*H = \Psi^*H = G \circ \Psi(\theta, p, t, E) - E = H \circ \Phi + \widetilde{E}(\theta, p, t). \tag{2.2}$$

The Aubry, Mather and Mañé sets are invariant under exact symplectic coordinate change in the following sense.

Proposition 2.7 ([10], [67]). *Suppose H and Φ^*H are Tonelli, and let Ψ be the extension of Φ. Let $(c, \alpha) \in \mathbb{R}^n \times \mathbb{R} \simeq H^1(\mathbb{T}^n \times \mathbb{T}, \mathbb{R})$, and let $\Psi^*(c, \alpha) = (c^*, \alpha^*)$ be the push forward of the cohomology class via the identification $H^1(\mathbb{T}^n \times \mathbb{T}, \mathbb{R}) \simeq H^1(\mathbb{T}^n \times \mathbb{R}^n \times \mathbb{T} \times \mathbb{R}, \mathbb{R})$. Then*

$$\alpha = \alpha_H(c) \quad \Longleftrightarrow \quad \alpha^* = \alpha_{\Phi^*H}(c^*).$$

Let us denote

$$(\Phi_H^*c, \alpha^*) = \Psi^*(c, \alpha_H(c)).$$

Then

$$\widetilde{\mathcal{M}}_H(c) = \Phi\left(\widetilde{\mathcal{M}}_{\Phi^*H}(\Phi_H^*c)\right), \quad \widetilde{\mathcal{A}}_H(c) = \Phi\left(\widetilde{\mathcal{A}}_{\Phi^*H}(\Phi_H^*c)\right),$$

$$\widetilde{\mathcal{N}}_H(c) = \Phi\left(\widetilde{\mathcal{N}}_{\Phi^*H}(\Phi_H^*c)\right).$$

Note in the particular case when Φ is homotopic to identity, $\Phi_H^* c = c$ and $\alpha_{\Phi^* H}(c) = \alpha_H(c)$.

Lemma 2.8. *Let Φ be an exact symplectic coordinate change on $\mathbb{T}^n \times \mathbb{R}^n \times \mathbb{T}$. Then tuple (H, c) satisfies (**Bif**) or (**Ar**) if and only if $(\Phi^* H, \Phi_H^* c)$ satisfies the same conditions. Suppose in addition that $\widetilde{\mathcal{A}}_H^0(c) = \widetilde{\mathcal{N}}_H^0(c)$. Then (H, c) satisfies (**Ma**) if and only if $(\Phi^* H, \Phi_H^* c)$ satisfies the same conditions.*

Proof. Since our symplectic coordinate changes are always identity in the t components, the invariance of Aubry and Mañé sets implied the invariance of their zero section under the map $\Phi(\cdot, \cdot, 0)$. The invariance of (**Bif**) follows. For the invariance of (**Ma**), note that due to the graph property, $\widetilde{\mathcal{A}}_H^0(c)$ is contractible in $\mathbb{T}^n \times \mathbb{R}^n$ if and only if $\mathcal{A}_H^0(c)$ is contractible in $\mathbb{T}^n$. Therefore the contractibility of Aubry set is invariant, and since the Aubry set coincides with the Mañé set by assumption, (**Ma**) is invariant.

For (**Ar**), let Ψ be the extension of Φ as in (2.1), and let us extend Ξ trivially to $\mathbb{T}^n \times \mathbb{R}^n \times \mathbb{T}$ or $\mathbb{T}^n \times \mathbb{R}^n \times \mathbb{R}$ without changing its name. Let Φ_1 be an exact symplectic change homotopic to Φ, with extension Ψ_1, such that

$$\Phi \circ \Xi = \Xi \circ \Phi_1, \quad \Psi \circ \Xi = \Xi \circ \Psi_1.$$

Note that $\Xi^* c$, defined as the push forward of $H^1(\mathbb{T}^n \times \mathbb{R}^n \times \mathbb{T} \times \mathbb{R}, \mathbb{R})$ under the identification with $H^1(\mathbb{T}^n \times \mathbb{T}, \mathbb{R})$, is identical to $\xi^* c$. We have

$$\Psi_1^*(\Xi^* c, \alpha_{H \circ \Xi}(\Xi^* c)) = \Psi_1^* \Xi^*(c, \alpha_H(c)) = \Xi^* \Psi^*(c, \alpha_H(c)).$$

Since Ξ is independent of t, Ξ^* is identity in the last component. We conclude that $(\Phi_1)_{H \circ \Xi}^* \Xi^* c = \Xi^*(\Phi_H^* c)$. Moreover,

$$(\Phi \circ \Xi)^* H = (\Phi^* H) \circ \Xi, \quad (\Xi \circ \Phi_1)^* H = \Phi_1^*(H \circ \Xi).$$

Then

$$\Phi_1 \left(\widetilde{\mathcal{N}}_{(\Phi \circ \Xi)^* H}((\Phi \circ \Xi)_H^* c) \right) = \Phi_1 \left(\widetilde{\mathcal{N}}_{\Phi_1^*(H \circ \Xi)}((\Phi_1)_{H \circ \Xi}^* \Xi^* c) \right) = \widetilde{\mathcal{N}}_{H \circ \Xi}(\Xi^* c)$$

and

$$\Phi_1 \left(\Xi^{-1} \widetilde{\mathcal{N}}_{\Phi^* H}(\Phi_H^* c) \right) = \Xi^{-1} \Phi \left(\Xi^{-1} \widetilde{\mathcal{N}}_{\Phi^* H}(\Phi_H^* c) \right) = \Xi^{-1} \left(\widetilde{\mathcal{N}}_H(c) \right),$$

therefore

$$\Phi_1 \left(\widetilde{\mathcal{N}}_{(\Phi^* H) \circ \Xi}(\Xi^*(\Phi_H^* c)) \setminus \Xi^{-1} \widetilde{\mathcal{N}}_{\Phi^* H}(\Phi_H^* c) \right) = \widetilde{\mathcal{N}}_{H \circ \Xi}(\Xi^* c) \setminus \Xi^{-1} \widetilde{\mathcal{N}}_H(c).$$

This implies invariance of (**Ar**) after considering the time-zero section of the above equality. $\qquad\square$

Our definition of exact symplectic coordinate change for a time-periodic system is somewhat restrictive, and in particular, it does not apply directly to the linear coordinate change performed at the double resonance. In that setting, we will prove invariance of the Mather, Aubry, and Mañé sets directly.

2.4 NORMAL HYPERBOLICITY AND AUBRY-MATHER TYPE

A 2-dymensional normally hyperbolic invariant cylinder is called *symplectic* if the restriction of the canonical form $d\theta \wedge dp$ is non-degenerate on its tangent space. Loosely speaking, a pair (H_*, c_*) is called of *Aubry-Mather type* (AM type for short, refer to Definition 8.1 for details) if:

1. The discrete Aubry set $\widetilde{\mathcal{A}}^0_{H_*}(c_*)$ is contained in a *2-dimensional* normally hyperbolic invariant cylinder on which the restriction of the symplectic form is non-degenerate.
2. There is $\sigma > 0$ such that the following holds for $c \in B_\sigma(c_*)$ and $H \in \mathcal{V}_\sigma(H_*)$:
   a) The discrete Aubry set satisfies the graph property under the local coordinates of the cylinder.
   b) When the Aubry set is an invariant graph, then locally the unstable manifold of the Aubry set is a graph over the configuration space $\mathbb{T}^2$.

This definition gives an abstract version of the setting seen in the *a priori unstable* systems.

Theorem 2.9 (See Theorem 8.6). *Suppose $H_* \in C^r$, $r \geq 2$, (H_*, c_*) is of Aubry-Mather type, and $\Gamma \subset \mathbb{R}^2$ is a smooth curve containing c_* in the relative interior. Then there is $\sigma > 0$ such that for all $c \in \overline{B_\sigma(c_*)} \cap \Gamma$, the following dichotomy holds for a C^r-residual subset of $H \in \mathcal{V}_\sigma(H_*)$:*

1. *Either the projected Mañé set $\mathcal{N}^0_H(c)$ is contractible as a subset of $\mathbb{T}^2$ (Mather mechanism (**Ma**));*
2. *Or there is a finite covering map Ξ such that the set*

$$\widetilde{\mathcal{N}}^0_{H \circ \Xi}(\xi^* c) \setminus \Xi^{-1} \widetilde{\mathcal{N}}^0_H(c)$$

*is totally disconnected (Arnold mechanism (**Ar**)).*

We say (H_*, c_*) is of *bifurcation Aubry-Mather type* if there exist two normally hyperbolic invariant cylinders, such that the *local Aubry set* restricted to each cylinder satisfies the conditions of Aubry-Mather type. The precise definition is given in Definition 8.3.

We will also consider a particular (and simpler) bifurcation. We say (H_*, c_*) is of *asymmetric bifurcation type* if there exists one normally hyperbolic invariant cylinder and a hyperbolic periodic orbit, such that the Aubry set is either contained in the cylinder (and of Aubry-Mather type) or the periodic orbit, see Definition 8.5.

We state the consequence of these definitions in terms of diffusion.

Theorem 2.10 (See Theorem 8.7). *Suppose $H_* \in C^r$, $r \geq 2$, (H_*, c_*) is of bifurcation Aubry-Mather type or asymmetric bifurcation type, and $\Gamma \subset \mathbb{R}^2$ is a smooth curve containing c_* in the relative interior. Then there is $\sigma > 0$ and an open and dense subset $\mathcal{R} \subset \mathcal{V}_\sigma(H_*)$ such that for each $H \in \mathcal{R}$ and each $c \in \Gamma \cap B_\sigma(c_*)$, (**Bif**) holds on at most finitely many c's, and for all other c's either (**Ma**) or (**Ar**) holds.*

The following Proposition is a direct consequence of Theorem 2.9 and Theorem 2.10.

Proposition 2.11. *Suppose $\Gamma \subset B^2$ is a piecewise smooth curve of cohomologies such that for each $c \in \Gamma$, the pair (H_*, c) is of Aubry-Mather type, bifurcation Aubry-Mather type, or asymmetric bifurcation type. Then there is $\sigma > 0$ and a residual subset $\mathcal{R}_\sigma(H_*) \subset \mathcal{V}_\sigma(H_*)$, such that either (**Ma**), (**Bif**) or (**Ar**) holds for each $H \in \mathcal{R}_\sigma(H_*)$ and each $c \in \Gamma$.*

Proof. For a piecewise smooth $\Gamma = \bigcup_{i=1}^m \Gamma_i$, we can extend each Γ_i to Γ_i' smoothly, such that Γ_i is contained in the relative interior of Γ_i'. We then apply Theorem 2.9 and Theorem 2.10 to each $c \in \Gamma_i$ relative to the smooth curve Γ_i', to get the conclusion of our Proposition for $c' \in B_{\sigma(c)}(c)$, and $H \in \mathcal{R}_{\sigma(c)}(H_*) \subset \mathcal{V}_{\sigma(c)}(H_*)$. The proposition then follows by considering a finite covering of Γ by $B_{\sigma(c_j)}(c_j)$, and taking finite intersection of residual subsets $\mathcal{R}_{\sigma(c_j)}(H_*)$. $\square$

We now describe the selection of cohomologies and prove AM type in each of the two regimes. Single resonance is covered in Chapter 3. Double resonance is split into two chapters: Chapter 4 covers the geometrical part, while Chapter 5 covers the variational part.

Chapter Three

Normal forms and cohomology classes at single resonances

3.1 RESONANT COMPONENT AND NON-DEGENERACY CONDITIONS

Let $(k_1, \Gamma_{k_1}) \in \mathcal{K}$ be a resonant segment in the diffusion path. Define the resonant component of H_1 relative to the single resonance k_1 as follows:

$$[H_1]_{k_1}(\theta, p, t) = \sum_{k \in k_1 \mathbb{Z}} h_k(p) e^{2\pi i k \cdot (\theta, t)},$$

where h_k are the Fourier coefficients of $H_1(\theta, p, t)$. Since $[H_1]_{k_1}$ only depends on the variables $k_1 \cdot (\theta, t)$ and p, we define a function $Z_{k_1} = Z_{k_1}(\theta^s, p) : \mathbb{T} \times \mathbb{R}^2 \to \mathbb{R}$ by the relation

$$Z_{k_1}(k_1 \cdot (\theta, t), p) = [H_1]_{k_1}(\theta, p, t).$$

For $p_0 \in \Gamma_k$, define the following conditions:

[$SR1_\lambda$] For all $p \in B_\lambda(p_0)$, the function $Z_{k_1}(\cdot, p)$ achieves a global maximum at $\theta_*^s(p) \in \mathbb{T}$, and for all $\theta^s \in \mathbb{T}$,

$$Z_{k_1}(\theta^s, p) - Z_{k_1}(\theta_*^s(p), p) < \lambda d(\theta^s, \theta_*^s(p))^2, \quad \forall \theta^s \in \mathbb{T}.$$

[$SR2_\lambda$] For all $p \in B_\lambda(p_0)$, there exist two local maxima $\theta_1^s(p)$ and $\theta_2^s(p)$ of the function $Z_{k_1}(., p)$ in $\mathbb{T}$ satisfying

$$\partial_{\theta^s}^2 Z_{k_1}(\theta_1^s(p), p) < \lambda I, \quad \partial_{\theta^s}^2 Z_{k_1}(\theta_2^s(p), p) < \lambda I$$

and for all $\theta^s \in \mathbb{T}$,

$$Z_{k_1}(\theta^s, p)$$
$$< \max\{Z_{k_1}(\theta_1^s(p), p), Z_{k_1}(\theta_2^s(p), p)\} - \lambda \big(\min\{d(\theta^s - \theta_1^s), d(\theta^s - \theta_2^s)\}\big)^2.$$

Definition 3.1. *For a given $\lambda > 0$, we say that H_1 satisfies the condition $SR(k_1, \Gamma_{k_1}, \lambda)$ if for each $p_0 \in \Gamma_{k_1}$, at least one of [$SR1_\lambda$] and [$SR2_\lambda$] holds for the function $Z_{k_1}(\theta^s, p)$.*

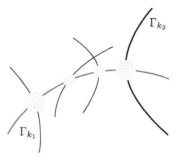

Figure 3.1: Single resonance after removing punctures

Proposition 3.2 (Proof is in Section 9.7). *The set of $H_1 \in \mathcal{S}^r$ such that $SR(k_1, \Gamma_{k_1}, \lambda)$ holds for some $\lambda > 0$ is open and dense.*

Let K be a large parameter, recall the strong additional resonances are defined by

$$\mathcal{K}^{\mathrm{st}}(k_1, \Gamma_{k_1}, K) = \{k_2 \in \mathbb{Z}_*^3 : |k_2| \le K, \ \Gamma_{k_1} \cap S_{k_2} \ne \emptyset\}.$$

We show generic forcing equivalence on each connected components of Γ_{k_1} minus $O(\sqrt{\epsilon})$-neighborhoods of the strong double resonances, called *punctures*. The following theorem is the main result of this section, and the proof is given in Section 3.3 assuming propositions proved in the later sections. For $M, K > 0$ denote

$$\Gamma_{k_1}^{SR}(M, K) := \Gamma_{k_1} \setminus \left(\bigcup_{k_2 \in \mathcal{K}^{\mathrm{st}}(k_1, \Gamma_{k_1}, K)} B_{2M\sqrt{\epsilon}}(\Gamma_{k_1, k_2}) \right). \qquad (3.1)$$

Theorem 3.3. *Suppose H_1 satisfies the condition $SR(k_1, \Gamma_{k_1}, \lambda)$ on Γ_{k_1}. Then there exist*

- $K, M, \epsilon_1 > 0$ *depending on* (D, k_1, λ), $\sigma = \sigma(k_1, H_0, \epsilon, H_1) > 0$,
- *for every* $0 < \epsilon < \epsilon_1$, *a residual subset* $\mathcal{R} \subset \mathcal{V}_\sigma(H_0 + \epsilon H_1)$,

such that the following holds for all $H \in \mathcal{R}$: for each $c \in \Gamma_{k_1}^{SR}(M, K)$ the associated Aubry or Mañé sets satisfy (**Ma**), (**Bif**), *or* (**Ar**). *As a result, each connected component of $\Gamma_{k_1}^{SR}(M, K)$ is contained in one forcing equivalent class.*

Refer to Figure 3.1 for an illustration.

3.2 NORMAL FORM

The classical partial averaging theory indicates that after a coordinate change, the system has the normal form $H_0 + \epsilon Z_{k_1} +$ h.o.t. away from punctures. In order to state the normal form, we need an anisotropic norm adapted to the perturbative nature of the system. Define

$$\|H_1(\theta, p, t)\|_{C_I^r} = \sup_{|\alpha| + |\beta| \le r} \epsilon^{\frac{|\beta|}{2}} \sup \left| \partial_{(\theta, t)}^\alpha \partial_p^\beta H_1(\theta, p, t) \right|, \qquad (3.2)$$

where $\alpha \in (\mathbb{Z}_+)^3, \beta \in (\mathbb{Z}_+)^2$ are multi-indices and $|\cdot|$ denotes the sum of the indices. The rescaled norm is similar to the C^r norm, but replace the p derivatives by the derivatives in $I = p/\sqrt{\epsilon}$, hence the name.

Theorem 3.4 (See end of this section). *Let $\kappa \in (0,1)$ there is $C = C(D, k_1, \kappa) > 1$ such that the following hold. Suppose $0 < \delta < C^{-1}\kappa$; set $K = C\delta^{-1}\kappa^{-1}$ and $M = \delta^{-2}$. Then for any $c \in \Gamma_{k_1}^{SR}(M, \delta^{-1})$ there exists $p_c \in \Gamma_{k_1}$ such that*

$$c \in B_{\kappa K \sqrt{\epsilon}}(p_c), \qquad (3.3)$$

and an C^∞ exact symplectic coordinate change homotopic to the identity

$$\Phi_\epsilon : \mathbb{T}^2 \times B_{K\sqrt{\epsilon}}(p_c) \times \mathbb{T} \to \mathbb{T}^2 \times B_{K\sqrt{\epsilon}/2}(p_c) \times \mathbb{T}$$

such that:

1.

$$(\Phi_\epsilon)^* H = H_0 + \epsilon[H_1]_{k_1} + \epsilon R, \qquad \|R\|_{C_I^2} \le C\delta. \qquad (3.4)$$

2. $\|\Pi_\theta(\Phi_\epsilon - Id)\|_{C_I^2} \le C\delta^2$ *and* $\|\Pi_p(\Phi_\epsilon - Id)\|_{C_I^2} \le C\delta^2\sqrt{\epsilon}.$

Here the C_I^2 norm is evaluated on the set $\mathbb{T}^2 \times B_{K\sqrt{\epsilon}/2}(p_c) \times \mathbb{T}$.

Remark 3.5.

- The C^∞ coordinate change is obtained by approximating a coordinate change that is only C^{r-1}, see Section 14.1. The reason this can be done is that we only need C^2 estimates of the coordinate change.
- When applying the theorem, κ will be chosen depending only on k_1, and in particular, independent of the "true small parameters" δ and ϵ.

We use the idea of Lochak (see for example [53]) to cover the action space with double resonances. A double resonance $p_0 = S_{k_1} \cap S_{k_2}$ corresponds to a periodic orbit of the unperturbed system H_0. More precisely, we have $\omega_0 = \Omega_0(p_0) := \nabla H_0(p_0)$ which satisfies $\mathbb{R}(\omega, 1) \cap \mathbb{Z}^3 \ne \emptyset$. Denote by $T_{\omega_0} = \min\{t > 0 : t(\omega_0, 1) \in \mathbb{Z}^3\}$ the minimal period.

The resonant lattice for p_0 is $\Lambda = \text{Span}_\mathbb{R}\{k_1, k_2\} \cap \mathbb{Z}^3$, and the resonant component is

$$[H_1]_{k_1,k_2} = \sum_{k \in \Lambda} h_k(p) e^{2\pi i k \cdot (\theta, t)}.$$

Proposition 3.6 (See Section 14.1). *Let $p_0 = \Gamma_{k_1,k_2}$, $T = T_{\omega_0(p_0)}$. Then for a parameter $C_1 > 1$, there exists $C = C(r, C_1) > 1$, $\epsilon_0 = \epsilon_0(r) > 0$ such that if $K_1 > C$ satisfies*

$$T < \frac{C_1}{K_1^2 \sqrt{\epsilon}},$$

then for each $0 < \epsilon < \epsilon_0$ there exists a C^∞ exact symplectic map

$$\Phi : \mathbb{T}^2 \times B_{K_1\sqrt{\epsilon}/2} \times \mathbb{T} \to \mathbb{T}^2 \times B_{K_1\sqrt{\epsilon}} \times \mathbb{T}$$

such that

$$(\Phi)^* H_\epsilon = H_0 + \epsilon [H_1]_{k_1,k_2} + \epsilon R_1,$$

where

$$\|R_1\|_{C_I^2} \leq C K_1^{-1},$$

and

$$\|\Pi_\theta(\Phi - Id)\|_{C_I^2} \leq C K_1^{-2}, \quad \|\Pi_p(\Phi - Id)\|_{C_I^2} \leq C K_1^{-2}\sqrt{\epsilon}.$$

Denote $\Lambda_1 = \mathrm{Span}_{\mathbb{R}}\{k_1\} \cap \mathbb{Z}^3$ and $\Lambda_2 = \mathrm{Span}_{\mathbb{R}}\{k_1, k_2\} \cap \mathbb{Z}^3$.

Lemma 3.7. *There is an absolute constant $C > 0$ such that if*

$$\min\{|k| : k \in \Lambda_2 \setminus \Lambda_1\} \geq K > 0,$$

we have

$$\|[H_1]_{k_1,k_2} - [H_1]_{k_1}\|_{C^2} \leq C K^{-\frac{1}{2}}.$$

Proof. Note that there is an absolute constant $C > 0$ such that for each two dimensional lattice $\Lambda \subset \mathbb{Z}^3$, we have

$$\sum_{k \in \Lambda \setminus \{0\}} |k|^{-2-\frac{1}{2}} < C.$$

Indeed, we can bound the sum above using the integral

$$\int_{|z| \geq 1, \, z \in \mathrm{Span}_{\mathbb{R}}\Lambda} |z|^{-2-\frac{1}{2}} dz.$$

Using the fact that $\|h_k\|_{C^2} \leq |k|^{2-r}\|H_1\|_{C^r}$, we have

$$\|[H_1]_{k_1,k_2} - [H_1]\|_{C^2} \leq \sum_{k \in \Lambda_2 \setminus \Lambda_1} |k|^{2-r} \leq K^{-\frac{1}{2}} \sum_{k \in \Lambda_2 \setminus \{0\}} |k|^{2+\frac{1}{2}-r} < C K^{-\frac{1}{2}}.$$

$\square$

The following lemma is an easy consequence of the Dirichlet theorem (see [53]).

Lemma 3.8. *There is $C = C(D, k_1) > 0$ such that for each $Q_1 > 1$ and each $c \in S_{k_1}$, there is a double resonance p_0 with $T_{\omega_0} < CQ_1$, and $\|c - p_0\| < C(T_{\omega_0} Q_1)^{-1}$, where $\omega_0 = \nabla H_0(p_0)$.*

Proof of Theorem 3.4. Denote $\tau = \delta^{-2}\sqrt{\varepsilon}$ and let $C_1 = C_1(k_1)$ be the larger of the two constants in Lemmas 3.7 and 3.8. Apply Lemma 3.8 using the parameter $Q_1 = C_1 \tau^{-1}$, then each $c \in S_{k_1}$ is contained in the $\frac{\tau}{T(p_c)} \leq K^2 \sqrt{\varepsilon}$ neighborhood of a double resonance p_c, whose period $T(\omega_c)$ is at most $C_1 Q_1$, where $\omega_c = \nabla H_0(p_c)$. Since $T_{\omega_0} \geq 1$, we get a first estimate

$$\|p_c - c\| < C_1 Q_1^{-1} = \tau = \delta^{-2}\sqrt{\epsilon}.$$

In particular, this implies if c is in the set (3.1) with $M = \delta^{-2}$, we have

$$p_c \notin \mathcal{K}^{\mathrm{st}}(k_1, \Gamma_{k_1}, \delta^{-1}).$$

Suppose $k_2 \in \mathbb{Z}_*^3$ is such that $p_c = \Gamma_{k_1, k_2}$, and since $\Lambda_2 \setminus \Lambda_1$ (see Lemma 3.7) contains only vectors larger than δ^{-1}, we have $T_{\omega_c} \geq C_1^{-1}\delta^{-1}$. We get the improved estimate

$$\|p_c - c\| \leq C_1 (T_{\omega_c} Q_1)^{-1} = C_1 \frac{\delta^{-2}\sqrt{\epsilon}}{T_{\omega_c}} \leq C_1^2 \delta^{-1}\sqrt{\epsilon} = \kappa K \sqrt{\epsilon},$$

where we've set $K = C_1^2 \kappa^{-1}\delta^{-1}$. Moreover, from Lemma 3.7,

$$\|[H_1]_{k_1, k_2} - [H_1]_{k_1}\|_{C^2} < C_1 \delta = C_1 \kappa^{-1} K^{-1}.$$

Set $C_2 = C_1^2 \kappa^{-2}$, then $T_{\omega_c} \leq \frac{C_1}{\delta^{-2}\sqrt{\epsilon}} = \frac{C_1^3 \kappa^{-2}}{C_1^2 \kappa^{-2}\delta^{-2}\sqrt{\epsilon}} = \frac{C_2}{K^2\sqrt{\epsilon}}$, and therefore Proposition 3.6 applies with the parameters C_2 and K. We obtain, for a constant $C_3 = C_3(C_2, \kappa)$,

$$\|R_1\|_{C_I^2} \leq C_3 K^{-1}.$$

Therefore

$$H_\epsilon \circ \Phi = H_0 + \epsilon[H_1]_{k_1} + \epsilon R,$$

where

$$\|R\|_{C_I^2} = \|\epsilon([H_1]_{k_1, k_2} - [H_1]_{k_1}) + \epsilon R_1\|_{C_I^2}$$
$$\leq C_3 C_1 \kappa^{-1} K^{-1} + C_1 K^{-\frac{1}{2}} \leq 2C_1 K^{-\frac{1}{2}}$$

as long as $K > C_3 \kappa^{-2}$, which is ensured if $\delta < C_3^{-1}\kappa$. Moreover,

$$\|\Pi_\theta(\Phi - Id)\|_{C_I^2} \leq C_3 K^{-2} \leq C_3 \delta^2,$$

$$\|\Pi_p(\Phi - Id)\|_{C_I^2} \leq C_3 K^{-2}\sqrt{\epsilon} \leq C_3 \delta^2 \sqrt{\epsilon}.$$

The theorem follows by setting $C = C_3$. □

3.3   THE RESONANT COMPONENT

Using the fact that $k_1 = (k_1^1, k_1^0) \in \mathbb{Z}^2 \times \mathbb{Z}$ is space irreducible, there is $k_2 = (k_2^1, k_2^0)$ such that $B_0^T := \begin{bmatrix} k_1^1 & k_2^1 \end{bmatrix} \in SL(2, \mathbb{Z})$. (This is the only pace where the space irreducibility condition is used.) Define

$$B = \begin{bmatrix} (k_1)^T \\ (k_2)^T \\ [0 \quad 0 \quad 1] \end{bmatrix} \in SL(3, \mathbb{Z}),$$

and

$$\Phi_L(\theta, p, t) = (\theta^s, \theta^f, p^s, p^f, t), \quad \begin{bmatrix} \theta^s \\ \theta^f \\ t \end{bmatrix} = B \begin{bmatrix} \theta \\ t \end{bmatrix}, \quad \begin{bmatrix} p^s \\ p^f \end{bmatrix} = (B_0^T)^{-1} p.$$

Let us note that $\|B\|, \|B^{-1}\|$ depends only on k_1. One verifies that Φ_L is a linear exact symplectic coordinate change. Note that $\theta^s = k_1 \cdot (\theta, t)$ and $[H_1]_{k_1} \circ \Phi_L$ depend only on θ^s, p^s, p^f. Let us write

$$N_\epsilon = \Phi_L^*(\Phi_\epsilon^* H_\epsilon) = H_0(p^s, p^f) + \epsilon Z(\theta^s, p^s, p^f) + \epsilon R(\theta^s, \theta^f, p^s, p^f, t), \quad (3.5)$$

where we abused notation by keeping the name of H_0 and R after the coordinate change. Let us also abuse notation by writing $\theta = (\theta^s, \theta^f)$ and $p = (p^s, p^f)$. Note that N_ϵ is defined on the set

$$\mathbb{T}^2 \times B_{K_1\sqrt{\epsilon}}(p_1) \times \mathbb{T}, \quad \text{where} \quad K_1 = \frac{K}{2\|B^{-1}\|}, \quad p_1 = (B_0^T)^{-1} p_0$$

and the resonant segment Γ_{k_1} is represented by $\Gamma^s = \{p : \partial_{p^s} H_0 = 0\}$ in the new coordinates.

For $\epsilon, \delta > 0$ and $p_1 \in \mathbb{R}^2$, define:

$$\mathcal{R}(\epsilon, \delta, p_1, K_1)$$
$$= \left\{ N_\epsilon = H_0 + \epsilon Z(\theta^s, p) + \epsilon R(\theta, p, t) : \quad \|R\|_{C_I^2(\mathbb{T}^2 \times B_{K_1\sqrt{\epsilon}}(p_1) \times \mathbb{T})} < \delta \right\}.$$

This set include the Hamiltonians in local normal form (3.5) for which the non-resonant component is smaller than $\epsilon\delta$.

We show that if ϵ, δ is small enough, and $N_\epsilon \in \mathcal{R}(\epsilon, \delta, p_1, K_1)$, the system N_ϵ admits a 3-dimensional normally hyperbolic invariant cylinder of the type

$$(\theta^s, p^f) = (\Theta^s, P^s)(\theta^f, p^f, t).$$

The discrete-time Aubry set $\widetilde{\mathcal{A}}_{N_\epsilon}^0$ is a graph over the θ^f component. The details will be given in Chapter 9, here we state the consequences of those results:

Proposition 3.9 (See Theorem 9.3). *Assume that $Z(\theta^s, p)$ satisfies condition $[SR1_\lambda]$ at $p_1 \in \Gamma^s$. Then there is $\delta_0, \epsilon_0, C > 0$ depending on D, λ such that if $K_1 > C$, $0 < \epsilon < \epsilon_0$, and $0 < \delta < \delta_0$, for each $N \in \mathcal{R}(\epsilon, \delta, p_1)$ and each $c \in B_{K_1\sqrt{\epsilon}/2}(p_1) \cap \Gamma^s$ the pair (N, c) is of Aubry-Mather type.*

Proposition 3.10 (See Theorem 9.4). *Consider N_ϵ as in Proposition 3.9, and assume that $Z(\theta^s, p)$ satisfies condition $[SR2_\lambda]$ at $p_2 \in \Gamma^s$. Then there is $\delta_0, \epsilon_0, C > 0$ depending on D, λ such that if $K_1 > C$, $0 < \epsilon < \epsilon_0$, and $0 < \delta < \delta_0$, there is an open and dense subset $\mathcal{R}_1 \subset \mathcal{R}(\epsilon, \delta, p_1, K_1)$, such that for each $c \in B_{K_1\sqrt{\epsilon}/2}(p_1) \cap \Gamma^s$ the pair (N, c) is of bifurcation Aubry-Mather type.*

Proof of Theorem 3.3. Denote $\mathcal{D}_r(p) = \mathbb{T}^2 \times B_r(p) \times \mathbb{T}$ for short. Let Γ be a connected component of (3.1) and $c_0 \in \Gamma$. Setting $\kappa = \frac{1}{8}(\max\{\|B\|, \|B^{-1}\|\})^{-2}$ which depends only on k_1, we apply Theorem 3.4 with parameters $\kappa, \delta > 0$ to obtain $C_0 = C_0(D, k_1) > 1$, $p_0 \in \Gamma_k$ such that $c_0 \in B_{\kappa K_0\sqrt{\epsilon}}(p_0)$, and the normal form transformation $\Phi_\epsilon : \mathcal{D}_{K_0\sqrt{\epsilon}/2}(p_0) \to \mathcal{D}_{K_0\sqrt{\epsilon}}(p_0)$. Let $K_1 = K/(2\|B^{-1}\|)$, $\Phi = \Phi_L \circ \Phi_\epsilon : \mathcal{D}_{K_1\sqrt{\epsilon}}(p_1) \to \mathcal{D}_{K_0\sqrt{\epsilon}}(p_0)$ and $N_\epsilon = \Phi^* H_\epsilon$. The mapping

$$\Phi^* : C^r(\mathcal{D}_{K_0\sqrt{\epsilon}}(p_0)) \to C^r(\mathcal{D}_{K_1\sqrt{\epsilon}}(p_1))$$

is a continuous linear mapping between Banach spaces, and we have $N_\epsilon \in \mathcal{R}(\epsilon, C_0\delta, p_1, K_1)$.

Under Φ, the correspondence between cohomolgies is given by the linear map $c \mapsto (B_0^T)^{-1}c$. Suppose c, p_0 are mapped to c_1, p_1; then we have $\|c_1 - p_1\| < \|B^{-1}\|\|c - p_0\| < K_0\sqrt{\epsilon}/(8\|B^{-1}\|) = K_1\sqrt{\epsilon}/4$, and hence $c_1 \in B_{K_1\sqrt{\epsilon}/4}(p_1) \cap \Gamma^s$. For ϵ_1 small enough, we choose δ sufficiently small such that Proposition 3.9 or 3.10 applies, depending on whether c_0 satisfies $[SR1_\lambda]$ or $[SR2_\lambda]$. It follows that for each $c \in B_{K_1\sqrt{\epsilon}/2}(p_1) \cap \Gamma^s$, (**Ma**), (**Bif**) or (**Ar**) holds on a C^r-residual subset $\mathcal{R}_\sigma(N_\epsilon)$ of $N \in \mathcal{V}_\sigma(N_\epsilon)$, in the space $C^r(\mathcal{D}_{K_1\sqrt{\epsilon}}(p_1))$.

By continuity of Φ^*, $(\Phi^*)^{-1}\mathcal{V}_\sigma(N_\epsilon)$ is an open set of $C^r(\mathcal{D}_{K_0\sqrt{\epsilon}}(p_0))$ containing H_ϵ, and therefore contains a neighborhood $\mathcal{V}_{\sigma'}(H_\epsilon)$ of H_ϵ. The subset $\mathcal{R}_{\sigma'}(H_\epsilon) = (\Phi^*)^{-1}\mathcal{R}_\sigma(N_\epsilon) \cap \mathcal{V}_{\sigma'}(H_\epsilon)$ is a residual subset of $\mathcal{V}_{\sigma'}(H_\epsilon)$. Applying Lemma 2.8 (invariance of diffusion mechanisms under symplectic coordinate changes), for each $H \in \mathcal{R}_{\sigma'}(H_\epsilon)$ and $c \in (B_0^T)B_{K_1\sqrt{\epsilon}/2}(p_1) \cap \Gamma^s \supset B_\sigma(c_0) \cap \Gamma$, one of (**Ma**), (**Bif**) or (**Ar**) hold.

We now apply the above argument to each $c \in \Gamma$, and establish (**Ma**), (**Bif**) or (**Ar**) for a neighborhood $B_{\sigma(c)}(c)$ of c, on a C^r residual subset $\mathcal{R}_c$ of $\mathcal{V}_{\sigma(c)}(H)$. By compactness, Γ can be covered by finitely many $B_{\sigma(c_i)}(c_i)$'s, then by taking intersections over $\mathcal{R}_{c_i}$, we conclude that our conditions hold on all $c \in \Gamma$, over a residual subset of $\mathcal{V}_{\sigma_0}(H_\epsilon)$, where $\sigma_0 = \min \sigma(c_i)$. The theorem follows. $\square$

Chapter Four

Double resonance: geometric description

In this chapter we describe the non-degeneracy condition and hyperbolic cylinders in the double-resonance regime. In the next chapter, we will return to variational setting, define the cohomology classes, and prove their forcing equivalence.

4.1 THE SLOW SYSTEM

Let $k \in \mathcal{K}^{\mathrm{st}}(k_1, \Gamma, K)$, and denote $p_0 = \Gamma_{k_1, k}$, and $\omega_0 = \nabla H_0(p_0)$. Define

$$\Lambda = \mathrm{Span}_{\mathbb{R}}\{k_1, k\} \cap \mathbb{Z}^3,$$

and choose $k_2 \in \mathbb{Z}^3_*$ such that $\Lambda = \mathrm{Span}_{\mathbb{Z}}\{k_1, k_2\}$. It is always possible to choose $|k_2| \leq |k_1| + K$.

Given $H_1 = \sum_{k \in \mathbb{Z}^3} h_k(p) e^{2\pi i k \cdot (\theta, t)}$, let

$$[H_1]_\Lambda = \sum_{k \in \Lambda} h_k(p) e^{2\pi i k \cdot (\theta, t)}.$$

After a symplectic coordinate change defined on the set $\mathbb{T}^2 \times B_{K\sqrt{\epsilon}}(p_0) \times \mathbb{T}$ (see Theorem 14.1), the system has the normal form:

$$N_\epsilon = \Phi_\epsilon^* H_\epsilon = H_0 + \epsilon[H_1]_\Lambda + O(\epsilon^{\frac{3}{2}}).$$

The system N_ϵ is conjugate to a perturbation of a two degrees of freedom mechanical system after a coordinate change and an energy reduction. The details are given in Chapter 14, here we give a brief description. Let $k_3 \in \mathbb{Z}^3$ be such that

$$B^T = \begin{bmatrix} k_1 & k_2 & k_3 \end{bmatrix} \in SL(3, \mathbb{Z}).$$

To define a symplectic coordinate change, we consider the corresponding autonomous system $N_\epsilon(\theta, p, t) + E$, and consider the coordinate change

$$(\theta, p, t, E) = \Phi_L(\varphi, I, \tau, F),$$

$$\begin{bmatrix} \theta \\ t \end{bmatrix} = B^{-1} \begin{bmatrix} \varphi \\ \tau/\sqrt{\epsilon} \end{bmatrix}, \qquad \begin{bmatrix} p - p_0 \\ E + H_0(p_0) \end{bmatrix} = B^T \begin{bmatrix} \sqrt{\epsilon} I \\ \epsilon F \end{bmatrix}. \tag{4.1}$$

One checks that

$$\left(\mathbb{T}^2 \times \mathbb{R}^2 \times \mathbb{T} \times \mathbb{R}, \quad d\theta \wedge dp + dt \wedge dE\right)$$
$$\overset{\Phi_L}{\rightarrow} \left(\mathbb{T}^2 \times \mathbb{R}^2 \times \mathbb{T} \times \mathbb{R}, \quad \frac{1}{\sqrt{\epsilon}}(d\varphi \wedge dI + d\tau \wedge dF)\right)$$

is an exact symplectic coordinate change. The transformed Hamiltonian $(N_\epsilon + E) \circ \Phi_L$ is no longer Tonelli in the standard sense, however by using a standard energy reduction on the energy level 0, with τ as the new time takes the system to

$$\frac{1}{\beta}\left(K(I) - U_0(\varphi) + \sqrt{\epsilon}P(\varphi, I, \tau)\right), \quad \varphi \in \mathbb{T}^2, I \in \mathbb{R}^2, \tau \in \sqrt{\epsilon}\mathbb{T},$$

where

$$\beta = k_3 \cdot (\omega_0, 1), \quad K(I) = \frac{1}{2}\left(B_0 \partial_{pp}^2 H_0(p_0) B_0^T\right), \quad B_0 = \begin{bmatrix} k_1^T \\ k_2^T \end{bmatrix},$$

and

$$U(k_1 \cdot (\theta, t), k_2 \cdot (\theta, t)) = -[H_1]_\Lambda(\theta, p_0, t).$$

The system

$$H^s(\varphi, I) = K(I) - U(\varphi) = K - U \tag{4.2}$$

is called the *slow mechanical system* (or slow system for short) and also denote

$$H_\epsilon^s(\varphi, I, \tau) = \frac{1}{\beta}\left(K(I) - U_0(\varphi) + \sqrt{\epsilon}P(\varphi, I, \tau)\right). \tag{4.3}$$

The non-degeneracy conditions at the double resonance p_0 are stated in terms of H^s.

4.2 NON-DEGENERACY CONDITIONS FOR THE SLOW SYSTEM

We consider the (shifted) energy as a parameter. For each $E > 0$, by the Maupertuis principle, the Hamiltonian dynamics on the energy surface $\mathcal{S}_E = \{(\varphi, I) : H^s(\varphi, I) + \min U(\varphi) = E\}$ is a reparametrization of the geodesic flow for the Jacobi metric

$$g_E(\varphi)(v) = 2(E + U(\varphi)) K^{-1}(v),$$

where $K^{-1}(v) = \frac{1}{2}\left(\partial_{II}^2 K\right)^{-1} v \cdot v$ is the Lagrangian associated to the Hamiltonian $K(I)$.

We will be interested in a special homology class $h = (0,1) \in \mathbb{Z}^2 \simeq H_1(\mathbb{T}^2, \mathbb{Z})$.

It represents orbits of the original system satisfying $k_1 \cdot (\dot\theta, 1) \simeq 0$, i.e. orbits that travel close to the resonance Γ_{k_1}. We assume the following non-degeneracy conditions:

[$DR1^h$] For each $E \in (0, \infty)$, each shortest closed geodesic (called a loop) of g_E in the homology class h is a hyperbolic orbit of the geodesic flow.

[$DR2^h$] At all but finitely many bifurcation values, there is only one g_E-shortest loop. At each bifurcation value E, there are exactly two shortest g_E loop denoted γ_h^E and $\bar\gamma_h^E$.

[$DR3^h$] At bifurcation value E_*,

$$\frac{d(\ell_E(\gamma_h^E))}{dE}\Big|_{E=E^*} \neq \frac{d(\ell_E(\bar\gamma_h^E))}{dE}\Big|_{E=E^*},$$

where l_E denote the g_E length of a loop.

We now discuss the conditions at the critical shifted energy $E = 0$. The Jacobi metric g_0 becomes degenerate at one point $\varphi_* = \mathrm{argmin}_\varphi U(\varphi)$. By performing a translation, we may assume $\varphi_* = 0$. While g_0 is no longer a Riemannian metric, it still makes sense to talk about shortest geodesics in the class h. Let γ_h^0 be one such shortest loops. Consider the following cases:

1. $0 \in \gamma_h^0$ and γ_h^0 is not self-intersecting. Call such homology class h *simple critical* and the corresponding geodesic γ_h^0 *simple loop*.
2. $0 \in \gamma_h^0$ and γ_h^0 is self-intersecting. Call such homology class h *non-simple* and the corresponding geodesic γ_h^0 *non-simple*.
3. $0 \notin \gamma_h^0$, then γ_h^0 is a regular geodesic. Call such homology class h *simple non-critical*.

Mather [65] proved that generically only these three cases occur (see the text that follows for the precise claim).

Lemma 4.1. *Let h be a non-simple homology class. Then for a generic potential U the curve γ_h^0 is the concatenation of two simple loops, possibly with multiplicities. More precisely, given $h \in H_1(\mathbb{T}^2, \mathbb{Z})$ generically there are simple homology classes h_1, $h_2 \in H_1(\mathbb{T}^2, \mathbb{Z})$ and integers n_1, $n_2 \in \mathbb{Z}_+$ such that the corresponding minimal geodesics $\gamma_{h_1}^0$ and $\gamma_{h_2}^0$ are simple and $h = n_1 h_1 + n_2 h_2$.*

We call γ_h^0 extensible if there exists a family of shortest curves γ_h^E converging to it in the Hausdorff topology. Consider the lift of γ_h^E to the universal cover $\mathbb{R}^2$, then as $E \to 0$ it converges to a periodic curve in $\mathbb{R}^2$ that consists of concatenation of $\gamma_{h_1}^0$ and $\gamma_{h_2}^0$. Suppose the order that the curves are concatenated is given by $\gamma_{h_{\sigma_1}}^0, \ldots, \gamma_{h_{\sigma_n}}^0$.

Lemma 4.2 (See Section 10.6). *Assume that γ_h^0 is extensible, then the sequence $(\sigma_n, \ldots, \sigma_n)$, as described, is uniquely determined up to cyclic permutation. We* *write*

$$\gamma_h^E \to \gamma_{h_{\sigma_1}}^0 * \cdots * \gamma_{h_{\sigma_n}}^0, \quad as \quad E \to 0.$$

We note that $(0,0)$ is a fixed point of the Hamiltonian flow $H^s(\varphi, I)$ which is hyperbolic if $\partial_{\theta\theta}^2 U(0) > 0$. Any simple g_0-shortest loop corresponds to a homoclinic orbit of the fixed point $(0,0)$. We impose the following non-degeneracy conditions:

[DR1^c] $(0,0)$ is a hyperbolic fixed point with distinct eigenvalues $-\lambda_2 < -\lambda_1 < 0 < \lambda_1 < \lambda_2$. Let $v_1^\pm, v_2^\pm$ be the eigendirections for $\pm\lambda_1, \pm\lambda_2$.

[DR2^c] There is a unique g_0-shortest loop in the homology h. If it is non-simple, then it is the concatenation of two simple loops $\gamma_{h_1}^0$ and $\gamma_{h_2}^0$.

[DR3^c] If γ_h^0 is simple critical, then it is *not* tangent to the $\langle v_2^+, v_2^- \rangle$ plane. If γ_h^0 is non-simple, then each of $\gamma_{h_1}^0$ and $\gamma_{h_2}^0$ is not tangent to the $\langle v_2^+, v_2^- \rangle$ plane.

[DR4^c]   – If γ_h^0 is simple non-critical, then γ_h^0 is hyperbolic.

      – If γ_h^0 is simple critical, then γ_h^0 is non-degenerate in the sense that it is the transversal intersection of the stable and unstable manifolds of $(0,0)$.

      – If γ_h^0 is non-simple, then each of $\gamma_{h_1}^0$ and $\gamma_{h_2}^0$ is non-degenerate.

The genericity of these conditions is summarized in the following statement.

Proposition 4.3. *The conditions* $[DR1^h - DR3^h]$ *and* $[DR1^c - DR4^c]$ *hold on an open and dense set of potentials* $U \in C^r(\mathbb{T}^2)$, *for* $r \geq 4$.

This proposition is the consequence of several statements given in the next section, namely Theorem 4.5, Proposition 4.6, and Proposition 4.7.

4.3 NORMALLY HYPERBOLIC CYLINDERS

Conditions $[DR1^h] - [DR3^h]$ ensures that for each $E_0 > 0$, the set

$$\bigcup_{E \in (E_0 - \delta, E_0 + \delta)} \eta_h^E$$

is a normally hyperbolic invariant cylinder. This cylinder does not necessarily extend to the shifted energy $E = 0$. The following statement ensures existence of cylinders near critical energy, using the conditions $[DR1^c] - [DR4^c]$.

Recall that γ_h^0 is a shortest curve of homology h in the critical energy. The corresponding Hamiltonian orbit is called η_h^0. Due to the symmetry of the system, we also have the shortest curve γ_{-h}^0 which coincides with γ_h^0 but has a different orientation.

Theorem 4.4 (See Chapter 10). *Suppose that* H^s *satisfies condition* $[DR1^c] - [DR4^c]$.

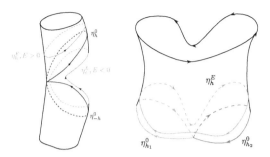

Figure 4.1: Extension of homoclinics to periodic orbits, simple and non-simple.

1. If γ_h^0 is simple, then there is $e > 0$ depending on H^s such that:

   a) For each $0 < E < e$, there exist periodic orbits η_h^E and η_{-h}^E, such that the projections $\gamma_h^E \to \gamma_h^0$ and $\gamma_{-h}^E \to \gamma_{-h}^0$ in the Hausdorff topology.

   b) For each $-e < E < 0$, there exists a periodic orbit η_c^E which shadows the concatenation of η_h^0 and η_{-h}^0.

   The union

   $$\bigcup_{0<E<e} \left(\eta_h^E \cup \eta_{-h}^E\right) \cup \eta_h^0 \cup \eta_{-h}^0 \cup \bigcup_{-e<E<0} \eta_c^E$$

   is a C^1 normally hyperbolic invariant manifold containing the homoclinics $\eta_{\pm h}^0$.

2. If γ_h^0 is non-simple: Let $\sigma_1, \ldots, \sigma_n$ be the sequence determined in Lemma 4.2. $[DR1^c] - [DR4^c]$ ensures γ_h^0 is extensible. More precisely, there is $e > 0$ such that for each $0 < E < e$, there is a periodic orbit γ_h^E such that $\gamma_h^E \to \gamma_{h_{\sigma_1}}^0 * \cdots * \gamma_{h_{\sigma_n}}^0$. Moreover, each γ_h^E is hyperbolic.

Theorem 4.4 allows us to prove Proposition 4.3 together with the following statements.

Theorem 4.5 (See Chapter 13). *Fix two parameters $0 < e_0 < \bar{E}$, then the set of potentials U such that $[DR1^h] - [DR3^h]$ holds on the smaller interval $E \in [e_0, \bar{E}]$ is open and dense in $C^r(\mathbb{T}^2)$, for $r \geq 3$.*

Proposition 4.6 (See Section 13.4). *The set of potentials $U \in C^r(\mathbb{T}^2)$ such that all γ_h^E for $E \geq \bar{E}$ is unique and hyperbolic is open and dense for $r \geq 4$.*

Proposition 4.7 (See Section 13.5). *The set of potentials U such that $[DR1^c] - [DR4^c]$ holds is open and dense in $C^r(\mathbb{T}^2)$, for $r \geq 2$.*

Proof of Proposition 4.3. Proposition 4.7 implies the set of potentials which satisfy $[DR1^c] - [DR4^c]$ is open and dense. By Theorem 4.4 the set of potentials $\mathcal{U}_{crit}$ such that there is $e_0 > 0$ such that all γ_h^E are unique and hyperbolic for all $0 < E < e_0$ is open and dense.

The set $\mathcal{U}_{High}$ of U's such that there exists $\bar{E} > 0$ such that all γ_h^E are unique and hyperbolic is open and dense by Proposition 4.6. By Theorem 4.5 the set of potentials $\mathcal{U}_{med}^{e_0,\bar{E}}$ such that for given $0 < e_0 < \bar{E}$, $[DR1^h] - [DR3^h]$ holds on $E \in [e_0, E]$ is also open and dense. As a result the set of potentials, where $[DR1^h] - [DR3^h]$ holds on $E \in (0, \infty)$ is

$$\mathcal{U}_{crit} \cap \bigcup_{0 < e_0 < \bar{E}} \mathcal{U}_{med}^{e_0,\bar{E}} \cap \mathcal{U}_{high},$$

which is open and dense.   □

Diffusion across a double resonance: a geometric description

The diffusion across a double resonance may be described heuristically as follows:

- If h is a simple homology, then the cylinder extends to the shifted energy $E < 0$ and connects with the homology $-h$. As a result, the family of periodic orbits η_h^E, $E \geq 0$ and η_{-h}^E, $E \geq 0$ are all contained in a normally hyperbolic invariant manifold. This corresponds to a continuous curve of cohomologies that are of Aubry-Mather type. Moreover, this picture survives small perturbations of H^s.
- If h is non-simple, then the cylinder "pinches" at $E = 0$. In particular, after considering the perturbation $H^s + \sqrt{\epsilon}P$ of the slow system, the cylinder may not survive the perturbation for E sufficiently close to 0. However, $h = n_1 h_1 + n_2 h_2$, and for the simple homology h_1 (or h_2), there exists a simple cylinder due to Theorem 4.4, item 1. The two simple cylinders are tangent to the weak stable/unstable directions plane at the fixed point $(0,0)$. See Figure 4.2.

  To diffuse across a double resonance, we "jump" from the cylinder for homology h to the cylinder with homology h_1, then diffuse across to homology $-h_1$ since h_1 is now simple, then jump back to homology $-h$. All of these are realized by choosing the appropriate cohomology curves lie on these cylinders. This construction is detailed in Chapter 5. See also Figure 5.1.

4.4   LOCAL MAPS AND GLOBAL MAPS

In this section we outline the basic approach to prove Theorem 4.4, based on ideas of Shil'nikov and others [18, 73, 78]. The full proofs are given in Chapter 10.

Suppose h is a simple critical homology. Let $\eta^+ = \eta_h^0$ be the homoclinic orbit to the hyperbolic fixed point $O = (0,0)$, and $\eta^- = \eta_{-h}^0$ its time reversal. Condition $[DR2^c]$ ensures that $\eta^\pm$ are not tangent to the strong stable/unstable directions, which implies they must be tangent to the weak stable/unstable

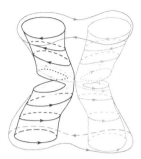

Figure 4.2: Hyperbolic invariant manifolds with kissing property

directions.

Consider four (3-dimensional) sections

$$\Sigma^u_\pm, \quad \Sigma^s_\pm$$

transverse to the weak stable and unstable eigen-directions, sufficiently close to the origin O, on each side of the equilibrium (see equation (10.5)). Up to renaming, we may assume that $\Sigma^{s/u}_+$ is transverse to η^+, and $\Sigma^{s/u}_-$ is transverse to η^-. We then define four local maps:

$$\Phi^{ij}_{\mathrm{loc}} : U_i (\subset \Sigma^s_i) \to \Sigma^u_j, \quad i, j \in \{+, -\}$$

as the Poincaré map between the corresponding sections. Note that these maps are *not defined on the whole section*, in particular, they are undefined along the (full) stable/unstable manifolds $W^{s/u}(O)$. However, they are defined on open sets. Moreover, we have a pair of global maps

$$\Phi^+_{\mathrm{glob}} : \Sigma^u_+ \to \Sigma^s_+, \quad \Phi^-_{\mathrm{glob}} : \Sigma^u_- \to \Sigma^s_-$$

which are the Poincaré maps along the orbits η^+ and η^-. These maps are well defined from a neighborhood of $\Sigma^u_\pm \cap \eta^\pm$ to a neighborhood of $\Sigma^s_\pm \cap \eta^\pm$. See Figure 4.3.

The periodic orbits obtained in Theorem 4.4 correspond to the fixed points of compositions of local and global maps, when restricted to the suitable energy surfaces. More precisely:

- The orbits η^E_h, $E > 0$ correspond to the fixed point of $\Phi^+_{\mathrm{glob}} \circ \Phi^{++}_{\mathrm{loc}}|_{S_E}$, where S_E denotes the energy surface $\{H^s + \min U(\varphi) = E\}$, and similarly for η^E_{-h}.
- The orbits η^E_c, $E < 0$ correspond to the fixed point of $\Phi^+_{\mathrm{glob}} \circ \Phi^{-+}_{\mathrm{loc}} \circ \Phi^-_{\mathrm{glob}} \circ \Phi^{+-}_{\mathrm{loc}}|_{S_E}$.

In the non-simple case, we similarly consider two homoclinics $\eta_1 = \eta_{h_1}$ and

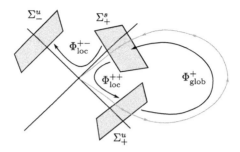

Figure 4.3: A local map and a global map

$\eta_2 = \eta_{h_2}$, and assume that both cross the sections $\Sigma_+^{s/u}$. Then we have the same local maps, and different global maps $\Phi_{\text{glob}}^{1/2}$. The periodic orbit η_h^E corresponds to the fixed point of

$$\prod_{i=n}^{1} \left(\Phi_{\text{glob}}^{\sigma_i} \circ \Phi_{\text{loc}}^{++} \right) |_{S_E},$$

where σ_i is the sequence in Lemma 4.2.

Chapter Five

Double resonance: forcing equivalence

5.1 CHOICE OF COHOMOLOGIES FOR THE SLOW SYSTEM

As in the case of single resonance, our strategy is to choose a continuous curve in the cohomology space and prove forcing equivalence up to a residual perturbation. To do this, we need to use the duality between homology and cohomology.

Let L be the Lagrangian associated to the Hamiltonian H, then the Euler-Lagrange flow

$$\dot{\theta} = v, \quad \frac{d}{dt}(\partial_v L) = \partial_x L$$

is conjugate to the Hamiltonian flow via

$$v = \partial_p H(\theta, p, t), \quad p = \partial_v L(\theta, v, t).$$

Let μ denote an invariant measure of the Euler-Lagrange flow. The rotation vector of μ is given by $\rho(\mu) = \int v \, d\mu(\theta, v, t)$. Then Mather's alpha and beta functions are defined as

$$\alpha_H(c) = -\inf_{\mu} \int (L(\theta, v, t) - c \cdot v) \, d\mu, \quad \beta_H(\rho) = \inf_{\rho(\mu)=\rho} \int L(\theta, v, t) d\mu.$$

A measure reaching the minimum in the definition of $\alpha_H(c)$ is called c-minimal. Then α and β are both convex and Fenchel dual of each other:

$$\beta_H(\rho) = \sup_{c \in \mathbb{R}^2} \{\rho \cdot c - \alpha_H(c)\}.$$

The Legendre-Fenchel transform of β is a set-valued function defined as

$$\mathcal{LF}_{\beta_H}(\rho) = \{c \in \mathbb{R}^2 : \quad \beta_H(\rho) = \rho \cdot c - \alpha_H(c)\}.$$

Geometrically,

$$\mathcal{LF}_{\beta_H}(\rho) = conv \, \{c : \text{there is an ergodic c-minimal } \mu \text{ such that } \rho(\mu) = \rho\},$$

where *conv* denotes the convex hull.

Let γ_h^E be a shortest loop for the Jacobi metric g_E. Let $T(\gamma_h^E)$ denotes its period under the Hamiltonian flow, and if γ_h^E is unique, we define

$$\lambda_h^E = 1/T(\gamma_h^E).$$

If H^s satisfies $[DR1^h] - [DR3^h]$, then there are at most finitely many E's (called bifurcation energy) such that there are two shortest loops γ_h^E and $\bar{\gamma}_h^E$. In this case, denote in addition $\bar{\lambda}_h^E = 1/T(\bar{\gamma}_h^E)$. We will show in Theorem 15.1 that the set $\mathcal{LF}_{\beta_{H^s}}(\lambda_h^E h) = \mathcal{LF}_{\beta_{H^s}}(\bar{\lambda}_h^E h)$, and therefore, the set $\mathcal{LF}_{\beta_H^s}(\lambda_h^E h)$ is independent of the choice of γ_h^E.

Each $\mathcal{LF}_{\beta_H^s}(\lambda_h^E h)$ is a segment of non-zero length parallel to $h^\perp$ and depends continuously on E. We call the union

$$\bigcup_{E>0} \mathcal{LF}_{\beta_{H^s}}(\lambda_h^E h) \tag{5.1}$$

the *channel* associated to the homology h, and we will choose a curve of co-homologies in the *interior* of this channel. The channel is connected at the bottom to the set $\mathcal{LF}_{\beta_H}(0)$, which has non-empty interior if H satisfies the condition $[DR1^c]$. The following proposition summarizes the channel picture and the relation to the Aubry sets.

Proposition 5.1 (See Chapter 15). *Assume that H^s satisfies the conditions $[DR1^h]$–$[DR3^h]$ and $[DR1^c]$–$[DR4^c]$. Then each $\mathcal{LF}_{\beta_{H^s}}(\lambda_h^E h)$ is a segment of non-zero length orthogonal to h, which varies continuously with respect to E.*

For $\bar{E} > 0$, let $\bar{c}_h : (0, \bar{E}] \to H^1(\mathbb{T}^2, \mathbb{R})$ be a C^1 function such that $\bar{c}_h(E)$ is in the relative interior of $\mathcal{LF}_{\beta_H}(\lambda_h^E h)$. The following hold.

1. *If E is not a bifurcation energy, then $\mathcal{A}_{H^s}(\bar{c}_h(E)) = \gamma_h^E$.*
2. *If E is a bifurcation energy, $\mathcal{A}_{H^s}(\bar{c}_h(E)) = \gamma_h^E \cup \bar{\gamma}_h^E$.*
3. *If h is simple, then the limit $\lim_{E \to 0} \mathcal{LF}_{\beta_{H^s}}(\lambda_h^E h)$ contains a segment of non-zero width. We assume, in addition, that $\bar{c}_h(0)$ is in the relative interior of this segment.*

   a) *If h is simple critical, then $\mathcal{A}_{H^s}(\bar{c}_h(0)) = \gamma_h^0$; for each $0 \le \lambda < 1$, we have $\mathcal{A}_{H^s}(\lambda \bar{c}_h(0)) = \{\varphi = 0\}$.*
   b) *If h is simple non-critical, then $\mathcal{A}_{H^s}(\bar{c}_h(0)) = \gamma_h^0 \cup \{\varphi = 0\}$; for each $0 \le \lambda < 1$, $\mathcal{A}_{H^s}(\lambda \bar{c}_h(0)) = \{\varphi = 0\}$.*
4. *If h is non-simple, then the limit $\lim_{E \to 0} \mathcal{LF}_{\beta_{H^s}}(\lambda_h^E h)$ is a single point.*

Note that due to symmetry of the system, $\mathcal{LF}_{\beta_H}(-\lambda h) = -\mathcal{LF}_{\beta_H}(\lambda h)$. Denote $\bar{c}_{-h}(E) = -\bar{c}_h(E)$. We now choose the cohomology classes for H^s as

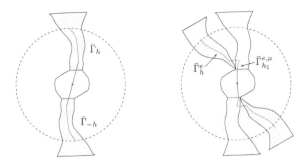

Figure 5.1: Choice of cohomology for H^s. Left: simple case; right: non-simple case.

follows: if h is simple (either critical or non-critical), we choose

$$\bar{\Gamma}_h = \bar{\Gamma}_h(\bar{E}) = \left(\bigcup_{0 \leq E \leq \bar{E}} \bar{c}_h(E) \right) \cup \left(\bigcup_{0 \leq \lambda \leq 1} \lambda \bar{c}_h(0) \right), \tag{5.2}$$

$$\bar{\Gamma}_h^{DR} = \bar{\Gamma}_h^{DR}(\bar{E}) = \bar{\Gamma}_h \cup \bar{\Gamma}_{-h} = \bar{\Gamma}_h \cup \left(-\bar{\Gamma}_h \right).$$

The curve $\bar{\Gamma}_h^{DR}$ is a continuous curve connecting $\bar{c}_h(\bar{E})$, 0, and $\bar{c}_{-h}(\bar{E})$.

If h is non-simple, then $h = n_1 h_1 + n_2 h_2$ is a combination of simple homologies h_1 and h_2. Let $0 < \mu < e$ be parameters. Define

$$\bar{\Gamma}_h^e = \bigcup_{e \leq E \leq \bar{E}} \bar{c}_h(E). \tag{5.3}$$

Since h_1 is simple critical, we choose a continuous curve $\bar{c}_{h_1}^\mu(E)$ such that

$$\|\bar{c}_{h_1}^\mu(0) - \bar{c}_h(0)\| < \mu.$$

We then define

$$\bar{\Gamma}_{h_1}^{e,\mu} = \left(\bigcup_{0 \leq E \leq e+\mu} \bar{c}_{h_1}^\mu(E) \right) \cup \left(\bigcup_{0 \leq \lambda \leq 1} \lambda \bar{c}_{h_1}(0) \right). \tag{5.4}$$

$\bar{\Gamma}_{h_1}^{e,\mu}$ is a continuous curve connecting $\bar{c}_{h_1}^\mu(e+\mu)$ with 0 (see Fig. 5.1, right). We then define

$$\bar{\Gamma}_h = \bar{\Gamma}_h^e \cup \bar{\Gamma}_{h_1}^{e,\mu}, \quad \bar{\Gamma}_h^{DR} = \bar{\Gamma}_h \cup \bar{\Gamma}_{-h}. \tag{5.5}$$

Let us note that the set $\bar{\Gamma}_h^{DR}$ consists of three connected components, $\bar{\Gamma}_h^e$, $\bar{\Gamma}_{-h}^e$, and $\bar{\Gamma}_{h_1}^{e,\mu} \cup \bar{\Gamma}_{-h_1}^{e,\mu}$.

5.2   AUBRY-MATHER TYPE AT A DOUBLE RESONANCE

Proposition 5.2. *Suppose* $H^s = K(I) - U(\varphi)$ *satisfies the non-degeneracy assumptions. Given* $C > 0$, *there is* $\epsilon_0, \delta > 0$ *depending on* H^s *and* C *such that for each* $0 < \epsilon < \epsilon_0$, *if* $U' \in \mathcal{V}_\delta(U)$ *and* $\|P\|_{C^2} < C$, *then for*

$$H_\epsilon^s = \frac{1}{\beta}\left(K(I) - U'(\varphi) + \sqrt{\epsilon}P(\varphi, I, \tau)\right),$$

each $\bar{c} \in \bar{\Gamma}_h^{DR}$ *is of one of four types: Aubry-Mather type, bifurcation Aubry-Mather type, asymmetric bifurcation type, or* $\widetilde{\mathcal{A}}_{H_\epsilon^s}$ *is a hyperbolic periodic orbit. Note that this applies to all types of homologies: simple critical, simple non-critical, and non-simple.*

Proof. We prove our proposition by referring to technical statements proved in later chapters.

(1) *Simple, critical homology.* Denote $c_0 = \bar{c}_h(0)$. Theorem 11.6 states that there exist $\epsilon_0, \epsilon, \delta > 0$ such that for all $0 < \epsilon < \epsilon_0$, $U' \in \mathcal{V}_\delta(U)$, and $c \in B_e(c_0)$, the pair H_ϵ^s, c is of Aubry-Mather type. The same theorem also states $H_\epsilon^s, \lambda c_0$ for $0 \leq \lambda \leq 1$ is of Aubry-Mather type. This covers the cohomologies

$$c \in \bigcup_{0 \leq E \leq e} \bar{c}_h(E) \cup \bigcup_{0 \leq \lambda \leq 1} \lambda \bar{c}_h(0).$$

The cohomologies $\bigcup_{e \leq E \leq \bar{E}} \bar{c}_h(E)$, are covered in Theorem 11.1, which states that for each $e > 0$, there is ϵ, δ as in our proposition, such that with respect to H_ϵ^s, the cohomology $\bar{c}_h(E)$, $E \geq e$ is of Aubry-Mather type if γ_h^E is the unique shortest curve, and of bifurcation Aubry-Mather type if there are two shortest curves. As a result, all cohomologies in $\bar{\Gamma}_h$ are of Aubry-Mather or bifurcation Aubry-Mather type, and by symmetry, the same holds for $\bar{\Gamma}_{-h}$. This proves our proposition in the simple homology case, see (5.2).

(2) *Non-simple homology.* In the non-simple case, the cohomology curve $\bar{\Gamma}_h$ is the disjoint union of two parts, namely

$$\bar{\Gamma}_h^e = \bigcup_{e \leq E \leq \bar{E}} \bar{c}_h(E), \quad \bar{\Gamma}_{h_1}^{e,\mu} = \left(\bigcup_{0 \leq E \leq e+\mu} \bar{c}_{h_1}^\mu(E)\right) \cup \left(\bigcup_{0 \leq \lambda \leq 1} \lambda \bar{c}_{h_1}(0)\right).$$

We note that h_1 is a simple homology and therefore each $c \in \bar{\Gamma}_{h_1}^{e,\mu}$ is of Aubry-Mather type in the same way as in case (1). On the other hand, each homology in $\bar{\Gamma}_h^e$ is of Aubry-Mather or bifurcation Aubry-Mather type since Theorem 11.1 applies the same way to simple and non-simple homology. We conclude that our Proposition holds in the non-simple case.

(3) *Simple, non-critical homology.* The high-energy case follows from Theorem 11.1 in the same way as before. The critical energy follows from Theorem 11.5, the main difference is that the critical energy is an asymmetric bifur-

cation (see Definition 8.5). For $c = \lambda \bar{c}_h(0)$ where $0 \leq \lambda < 1$, the Aubry set is a single periodic orbit as a perturbation of the hyperbolic fixed point $(0, 0)$. □

Corollary 5.3. *Suppose H^s_ϵ is from Proposition 5.2; then there is $\sigma > 0$ and a residual subset $\mathcal{R}$ of $V_\sigma(H^s_\epsilon)$, such that if $H' \in \mathcal{R}$, then each $\bar{c} \in \bar{\Gamma}^{DR}_h$ satisfies one of the diffusion mechanisms* (**Ma**), (**Bif**), *or* (**Ar**).

Proof. Note that Proposition 2.11 applies when c is of Aubry-Mather type, bifurcation Aubry-Mather type, or asymmetric bifurcation type. Then using Proposition 5.2, we only need to show the same conclusion holds if we add a fourth case, when the Aubry set is a single hyperbolic periodic orbit. However, in this case, $\tilde{\mathcal{N}}^0 = \tilde{\mathcal{A}}^0$ is discrete, so (**Ma**) applies. □

We now revert to the original coordinate system. Denote $p_0 = \Gamma_{k_1, k_2}$ the double resonance point. For a given cohomology class $\bar{c} \in \mathbb{R}^2$, we consider the pair c, α defined by the equality

$$\begin{bmatrix} c - p_0 \\ -\alpha + H_0(p_0) \end{bmatrix} = B^T \begin{bmatrix} \bar{c}\sqrt{\epsilon} \\ \alpha_{H^s_\epsilon}(\bar{c}) \epsilon \end{bmatrix}. \tag{5.6}$$

Then $\alpha = \alpha_{N_\epsilon}(c)$ (compare to (4.1)). If (5.6) is satisfied, we denote

$$(c, -\alpha) = \Phi^*_L(\bar{c}, \alpha_{H^s_\epsilon}(\bar{c})), \quad c = \Phi^*_{L, H^s_\epsilon}(\bar{c}).$$

Moreover, let us consider the autonomous version of the Aubry set:

$$\tilde{\mathcal{A}}_{H^s_\epsilon + F}(\bar{c}) = \{(\varphi, I, \tau, -H^s_\epsilon(\varphi, I, \tau)) : \quad (\varphi, I, \tau) \in \tilde{\mathcal{A}}_{H^s_\epsilon}(\bar{c})\},$$

and similarly define $\widetilde{\mathcal{M}}$ and $\widetilde{\mathcal{N}}$. It is proven in Proposition 14.9 that

$$\tilde{\mathcal{A}}_{N_\epsilon + E}(c) = \Phi_L\left(\tilde{\mathcal{A}}_{H^s_\epsilon + F}(\bar{c})\right), \quad \tilde{\mathcal{N}}_{N_\epsilon + E}(c) = \Phi_L\left(\tilde{\mathcal{N}}_{H^s_\epsilon + F}(\bar{c})\right). \tag{5.7}$$

Proposition 5.4. *Suppose $c, \bar{c}$ are related by (5.6). Then with respect to N_ϵ, c satisfies one of the diffusion mechanisms* (**Ma**), (**Bif**), *or* (**Ar**) *if and only if $\bar{c}$ does the same with respect to H^s_ϵ.*

Proof. Suppose N_ϵ, c and $H^s_\epsilon, \bar{c}$ satisfy our assumption. Consider the following subsets:

$$S = \{(\varphi, I, \tau, F) : H^\epsilon_s + F = 0\} \subset \mathbb{T}^2 \times \mathbb{R}^2 \times \sqrt{\epsilon}\mathbb{T} \times \mathbb{R}, \quad \Sigma_0 = S \cap \{\tau = 0\},$$

$$\hat{S} = \{(\theta, I, t, E) : N_\epsilon + E = 0\} \subset \mathbb{T}^2 \times \mathbb{R}^2 \times \mathbb{T} \times \mathbb{R}, \quad \hat{\Sigma}_0 = \hat{S} \cap \{t = 0\}.$$

Note that S is a graph over $\mathbb{T}^2 \times \mathbb{R}^2 \times \sqrt{\epsilon}\mathbb{T}$ and Σ_0 is a graph over $\mathbb{T}^2 \times \mathbb{R}^2$. We also have S is invariant under H^ϵ_s, $\tilde{\mathcal{A}}_{H^s_\epsilon + F}(\bar{c}) \subset S$, and Σ_0 is a global section of the flow on S. Suppose $(H^\epsilon_s, \bar{c})$ satisfies $\tilde{\mathcal{A}}_{H^s_\epsilon}(\bar{c}) = \tilde{\mathcal{N}}_{H^s_\epsilon}(\bar{c})$, and (**Ma**) applies.

This implies $\widetilde{\mathcal{A}}_{H^\epsilon_s+F}(\bar{c}) \cap \Sigma_0$ is a contractible subset of Σ_0. By (5.7),

$$\widetilde{\mathcal{A}}_{N_\epsilon+\epsilon}(c) = \Phi_L\left(\widetilde{\mathcal{A}}_{H^\epsilon_s+F}\right),$$

we get $\widetilde{\mathcal{A}}_{N_\epsilon+E}(c) \cap \Phi_L(\Sigma_0)$ is contractible in $\Phi_L(\Sigma_0)$. Since Φ_L is a conjugacy between Hamiltonian flows, $\Phi_L(\Sigma_0)$ is a global section for the flow of $N_\epsilon + E$ on the energy surface $\hat{S}$. Since $\hat{\Sigma}_0$ is also a global section, the Poincaré map $P : \Phi_L(\Sigma_0) \to \hat{\Sigma}_0$ is a diffeomorphism. As a result,

$$\widetilde{\mathcal{A}}_{H^\epsilon+E}(c) \cap \Sigma_0 = P\left(\widetilde{\mathcal{A}}_{N_\epsilon+E}(c) \cap \Phi_L(\Sigma_0)\right)$$

is contractible in $\hat{\Sigma}_0$. This proves (**Ma**) holds for (N_ϵ, c). The proofs for the converse and the other mechanisms are similar. $\qquad\square$

5.3   CONNECTING TO Γ_{K_1,K_2} AND $\Gamma^{SR}_{K_1}$

At this point it is natural to consider the cohomology class

$$\Phi^*_{L,H^s_\epsilon}\left(\bar{\Gamma}^{DR}_h\right) := \pi_c\left(\{\Phi^*_L(\bar{c}, \alpha_{H^s_\epsilon}(\bar{c})) : \quad \bar{c} \in \Gamma^{DR}_h\}\right) \tag{5.8}$$

(see (5.6)) where π_c denotes the projection $(c, -\alpha) \mapsto c \in \mathbb{R}^2$. We note that α is automatically determined by c via the relation $\alpha = \alpha_{N_\epsilon}(c) = \alpha_{H_\epsilon}(c)$.

We would like to choose the cohomology $\Phi^*_{L,H^s_\epsilon}\left(\bar{\Gamma}^{DR}_h\right)$ for the original system, but due to the ϵ dependence of the map, the new set does not necessarily contain the double resonance point Γ_{k_1,k_2}, nor does it connect to the cohomology $\Gamma^{SR}_{K_1}$ already chosen at single resonance. To solve this problem, we will add three pieces of "connectors" to the set: Γ^{con}_0 is used to connect to Γ_{k_1,k_2}, and $\Gamma^{con}_\pm$ to connect to $\Gamma^{SR}_{k_1}$. Then

$$\Gamma^{DR}_{k_1,k_2}(\epsilon, H_1) = \Phi^*_{L,H^s_\epsilon}\left(\bar{\Gamma}^{DR}_h\right) \cup \Gamma^{con}_0 \cup \Gamma^{con}_- \cup \Gamma^{con}_+. \tag{5.9}$$

See Figure 5.2.

5.3.1   Connecting to the double resonance point

We define

$$c_0 = \Gamma_{k_1,k_2}, \quad c^\epsilon_0 = \Phi^*_{L,H^s_\epsilon}(0, \alpha_{H^s_\epsilon}(0)), \quad \Gamma^{con}_0 = \bigcup_{s\in[0,1]}\{sc_0 + (1-s)c^\epsilon_0\}$$

and define h to be the homology class corresponding to k_1. Γ^{con}_0 is simply a line segment connecting Γ_{k_1,k_2} and c^ϵ.

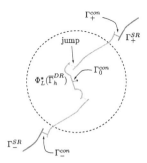

Figure 5.2: Cohomology curve at double resonance, with connectors.

Proposition 5.5. *Suppose H_1 satisfies the conditions $[DR1^h - DR3^h]$ and $[DR1^c - DR4^c]$ at the double resonance Γ_{k_1,k_2} relative to Γ_{k_1}. Then there is $\epsilon_0, \delta > 0$ depending only on H^s such that for $0 < \epsilon < \epsilon_0$, $H_1' \in \mathcal{V}_\delta(H_1)$, $0 \leq s \leq 1$, and $H_\epsilon' = H_0 + \epsilon H_1'$:*

$$\mathcal{N}_{H_\epsilon'}^0(sc_0 + (1-s)c_0^\epsilon) = \mathcal{A}_{H_\epsilon'}^0(sc_0 + (1-s)c_0^\epsilon) = \mathcal{A}_{H_\epsilon'}^0(c_0) = \mathcal{A}_{H_\epsilon'}^0(c_0^\epsilon)$$

is contractible in $\mathbb{T}^2$. As a result the cohomologies in Γ_0^{con} are forcing equivalent.

Proof. Consider $\bar{c}_0^\epsilon \in \mathbb{R}^2$, $\bar{\alpha}_0^\epsilon \in \mathbb{R}$ defined by the formula

$$\begin{bmatrix} \bar{c}_0^\epsilon \sqrt{\epsilon} \\ \bar{\alpha}_0^\epsilon \epsilon \end{bmatrix} = M^{-T} \begin{bmatrix} 0 \\ -\alpha_{H_\epsilon}(c_0) + H_0(c_0) \end{bmatrix}.$$

Then according to (5.6), $\Phi_{L,H_\epsilon^s}^*(\bar{c}_0, \bar{\alpha}_0) = c_0$. According to Lemma 7.9, we have $\|\alpha_{H_\epsilon}(c_0) - H_0(c_0)\| \leq C\epsilon$ for some C depending only on H_0; therefore $\|\bar{c}_0^\epsilon\| \leq C\sqrt{\epsilon}$.

Recall that $O = (0,0)$ is a hyperbolic fixed point of the system H^s, and $\widetilde{\mathcal{A}}_{H^s}(0) = O$. By standard perturbation theory of hyperbolic sets, there is a neighborhood $V \ni O$, such that the system $H_\epsilon^s = K - U' + \sqrt{\epsilon}P$ with $U' \in \mathcal{V}_\delta(U)$ and $0 < \epsilon < \epsilon_0$, H_ϵ^s admits a unique hyperbolic periodic orbit O_ϵ contained in V. Moreover, using the upper semi-continuity of the Aubry set Corollary 7.2, by possibly choosing δ and ϵ_0 smaller, we ensure for all $0 \leq \lambda \leq 1$, $\widetilde{\mathcal{A}}_{H_\epsilon^s}(\lambda \bar{c}_0^\epsilon) \subset V$, and therefore $\widetilde{\mathcal{A}}_{H_\epsilon^s}(\lambda \bar{c}_0^\epsilon) = \widetilde{\mathcal{A}}_{H_\epsilon^s}(0) = O_\epsilon$. In this case the Aubry set has a unique static class; hence the Aubry set coincides with the Mañé set, and also the discrete Aubry set is finite and therefore contractible. We now apply symplectic invariance (5.7) to get the same for the original system. $\square$

5.3.2 Connecting single and double resonance

The single-resonance cohomology $\Gamma_{k_1}^{SR}(M, K)$ is defined in (3.1). For each double resonance Γ_{k_1,k_2}, let $\Gamma_{k_1,k_2,\pm}^{SR}$ be the two connected components of (3.1) adjacent to Γ_{k_1,k_2}. We define connectors $\Gamma_{\pm}^{\mathrm{con}}$ which connect the double-resonance cohomology curve $\Phi_L^*\left(\bar{\Gamma}_h^{DR}\right)$ to $\Gamma_{k_1,k_2,\pm}^{SR}$, respectively.

Let $c \in \mathbb{R}^2 \simeq H^1(\mathbb{T}^2,\mathbb{R})$ be a cohomology, and let $\rho_H(c)$ denote the convex hull of rotation vectors of all c-minimal measures. This coincides with the Legendre-Fenchel transform relative to the alpha function

$$\rho_H(c) = \mathcal{LF}_{\alpha_H}(c) \subset \mathbb{R}^2.$$

Note that $\rho_H(c)$ is a set valued function taking values in convex sets. We have the following observations:

- Let $c \in \Gamma_{k_1}^{SR}$. We showed in Section 3.3 that after a normal form and linear coordinate change and turning H_ϵ to the normal form N_ϵ, the Aubry set $\widetilde{\mathcal{A}}_{N_\epsilon}(\Phi_L^* c)$ is a graph over the ϕ^f component. This implies the rotation vector $\rho_{N_\epsilon}(\Phi_L^* c)$ is parallel to $(0,1)$ (in (θ^s, θ^f) component). After reverting the coordinate change, this relation becomes

$$k_1 \cdot (\rho_{H_\epsilon}(c), 1) = 0.$$

- If $c \in \Gamma_h^{DR}$, let $\bar{c}$ be the associated cohomology for H_ϵ^s (via (5.6)). Since the cohomologies we chose are in the channel associated to $h = (0,1)$, $\rho_{H_\epsilon^s}(\bar{c})$ is parallel to h. By the same reasoning as in the single-resonance case, after reverting the coordinate change, we have

$$k_1 \cdot (\rho_{H_\epsilon}(c), 1) = 0.$$

We conclude that in both cases, the rotation vectors of the chosen cohomologies stay on the rational line

$$\Omega_{k_1} = \{\omega : \quad k_1 \cdot (\omega, 1) = 0\}.$$

Let us also denote $\Omega_{k_1,k_2} = \Omega_{k_1} \cap \Omega_{k_2}$, $c_0 = \Gamma_{k_1,k_2}$, and $\omega_0 = \Omega_{k_1,k_2}$.

The main observation is that the rotation vectors of the curves $\Gamma_{k_1}^{SR}(M, K)$ and Γ_{k_1,k_2}^{DR} are both contained in Ω_{k_1}, and their intersection is non-empty; see Figure 5.3. To prove this statement, we show that the rotation vector of $c \in \Gamma_{k_1}^{SR}(M, K)$ is $O(\sqrt{\epsilon})$ close to ω_0 at its nearest point, while the rotation vector of $c \in \Gamma_{k_1,k_2}^{DR}$ is $C^{-1}E\sqrt{\epsilon}$ away from ω_0 when E is sufficiently large. Since both sets are connected and lie on the same line, they have to intersect.

Proposition 5.6. *Let $\Gamma_{k_1,k_2,\pm}^{SR}$ and Γ_{k_1,k_2} be as before. Suppose H_1 satisfies the conditions $[DR1^h - DR3^h]$ and $[DR1^c - DR4^c]$ relative to Γ_{k_1}. Then there is $\epsilon_0, \delta > 0$ depending only on H^s such that such that if $0 < \epsilon < \epsilon_0$, and $H_1' \in \mathcal{V}_\delta(H_1)$, for $H_\epsilon' = H_0 + \epsilon H_1'$:*

1. *For each $c \in \Gamma_{k_1}^{SR}(M, K)$, $\rho_{H_\epsilon'}(c)$ is a single point contained in Ω_{k_1}. The*

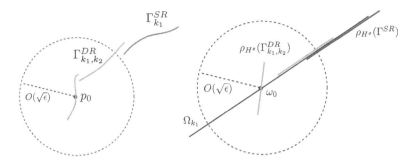

Figure 5.3: Left: cohomology curves; right: rotation vectors.

function $\rho_{H_\epsilon}(c)$ is continuous on Γ^{SR}, and there is $C > 0$ such that

$$\|\rho_{H_\epsilon}(c) - \omega_0\| \leq C\|c - c_0\|, \quad \forall\, c \in \Gamma_{k_1}^{SR}(M, K).$$

2. *Let $c_h(E) = \Phi_L^*\left(\bar{c}_h(E)\right) \in \Gamma_{k_1,k_2}^{DR}$. Then each $\rho_{H_\epsilon'}(c_h(E))$ is a single point contained in Ω_{k_1}, and there are $\check{C}, E_0 > 0$ such that*

$$\|\rho_{H_\epsilon'}(c_h(E)) - \omega_0\| \geq \check{C}^{-1} E\sqrt{\epsilon}, \quad E \geq E_0.$$

As a result, there exists $\bar{E} > 0$ such that for $E > \bar{E}$, the rotation vectors $\rho_{H_\epsilon'}(c_h(E))$ coincide with one of $\rho_{H_\epsilon'}(\Gamma_{k_1,k_2,\pm}^{SR})$, see Figure 5.3. Similarly, for E sufficiently large, the rotation vectors $\rho_{H_\epsilon'}(c_{-h}(E))$ coincide with one of $\rho_{H_\epsilon'}(\Gamma_{k_1,k_2,\pm}^{SR})$ (which are not covered in the first case).

Proof. In the single-resonance regime, after a linear coordinate change (Section 3.3) the system is converted to $H_0 + \epsilon Z(\theta^s, p) + \epsilon R(\theta^s, \theta^f, p^s, p^f, t)$, where the invariant cylinder is given by $(\theta^s, p^s) = (\Theta^s, P^s)(\theta^f, p^f, t)$, and the Aubry set for any $c \in \Gamma^s = \{(\partial_p H_0(0), 1) \cdot (1, 0, 0) = 0\}$ is a graph over (θ^f, t) (see Theorem 9.1 and Theorem 9.2) This implies that any c admits a unique rotation vector, as different rotation vectors will result in intersecting minimizing orbits, which violates the graph theorem. By reverting the coordinate change, we also obtain $\rho_{H_\epsilon}(c) \in \Omega_{k_1}$. We then apply Corollary 7.8, to get (in general)

$$\|\rho_{H_\epsilon}(c) - \nabla H_0(c)\| \leq C\sqrt{\epsilon},$$

while

$$\|\nabla H_0(c) - \omega_0\| = \|\nabla H_0(c) - \nabla H_0(c_0)\| \leq C\|c - c_0\|$$

and recall that $\|c - c_0\| \geq K\sqrt{\epsilon}$ for any $c \in \Gamma_{k_1}^{SR}$, item (1) follows.

In the double-resonance case, the cohomology curves $\bar{c}_h(E)$ at high-energy are of Aubry-Mather type corresponding to the homology class $h = (1, 0)$. The same arguments as in the single-resonance case imply $\rho_{H_\epsilon^s}(c)$ is unique, and contained in the line $\omega_1 = 0$. Again by reverting the coordinate change, we obtain the desired property for $\rho_{H-\epsilon}(c)$. For the inequality, note that if any point (φ, I, t) in the Aubry set $\widetilde{\mathcal{A}}_{H_\epsilon^s}(c)$ satisfies $H^s(\varphi, I) \geq E_0$, we have $K(I) \geq H^s(\varphi, I) - \|U\|_{C^0} \geq \frac{1}{2}H^s(\varphi, I)$ if $E_0 \geq 2\|U\|_{C^0}$. As a result

$$\min_{(\varphi, I, \tau) \in \widetilde{\mathcal{A}}_{H_\epsilon^s}(\bar{c}_h(E))} \|I\| \geq C^{-1} \qquad \min_{(\varphi, I, \tau) \in \widetilde{\mathcal{A}}_{H_\epsilon^s}(\bar{c}_h(E))} \sqrt{K(I)} \geq C^{-1}\sqrt{E}$$

for a suitable $C > 1$. Reverting the coordinate change Φ_L implies

$$\min_{(\theta, p, t) \in \widetilde{\mathcal{A}}_{H_\epsilon}(c_h(E))} \|p - p_0\| \geq C^{-1}\|I\|\sqrt{\epsilon} \geq C^{-1}\sqrt{E}\sqrt{\epsilon}.$$

Finally, we apply Corollary 7.8 again to get our inequality.   □

We have the following lemma:

Lemma 5.7 (Proposition 6 of [56]). *Let $c_1, c_2 \in H^1(\mathbb{T}^2, \mathbb{R})$ and $\rho \in H_1(\mathbb{T}^2, \mathbb{R})$ satisfy $\rho_H(c_1) = \rho_H(c_2) = \rho$, and both c_1, c_2 lie in the relative interior of $\mathcal{LF}_{\beta_H}(\rho)$. Then $\widetilde{\mathcal{A}}_H(c_1) = \widetilde{\mathcal{A}}_H(c_2)$.*

Proposition 5.8. *Suppose H_1 satisfies the conditions $[DR1^h - DR3^h]$ and $[DR1^c - DR4^c]$ relative to Γ_{k_1}. Then there is $\epsilon_0, \delta > 0$ depending only on H^s such that for $0 < \epsilon < \epsilon_0$ and $H_1' \in \mathcal{R}_\delta(H_1)$, there is $c_1^\pm \in \Gamma_{k_1}^{SR}(M, K)$ and $c_2^\pm \in \Gamma_{k_1, k_2}^{DR}$, such that for all $0 \leq \lambda \leq 1$ and $H_\epsilon' = H_0 + \epsilon H_1'$*

$$\mathcal{N}_{H_\epsilon'}^0(\lambda c_1^\pm + (1 - \lambda)c_2^\pm) = \mathcal{A}_{H_\epsilon'}^0(\lambda c_1^\pm + (1 - \lambda)c_2^\pm) = \mathcal{A}_{H_\epsilon'}^0(c_1^\pm) = \mathcal{A}_{H_\epsilon'}^0(c_2^\pm)$$

is contractible in $\mathbb{T}^2$. As a result both cohomology curves

$$\Gamma_+^{\mathrm{con}} = \bigcup_{s \in [0,1]} \{sc_1^+ + (1 - s)c_2^+\}, \qquad \Gamma_-^{\mathrm{con}} = \bigcup_{s \in [0,1]} \{sc_1^- + (1 - s)c_2^-\}$$

are contained in a single forcing equivalent class.

Proof. Proposition 5.6 implies the curves $\rho_{H_\epsilon}(\Gamma_{k_1}^{SR})$ and $\rho_{H_\epsilon}(\Gamma_{k_1, k_2}^{DR})$ overlap on an interval contained in Ω_{k_1}, for all $H_1' \in \mathcal{V}_\delta(H_1)$. In particular, there must be $c_1 \in \Gamma_{k_1}^{SR}$ and $c_2 \in \Gamma_{k_1, k_2}^{DR}$ where they both have rational rotation vectors. We now assume that $H_\epsilon' = H_0 + \epsilon H_1'$ satisfies the residual condition that all Aubry sets with rational rotation vector are supported on a hyperbolic periodic orbit, in this setting $\mathcal{A} = \mathcal{N}$, c_1, c_2 are contained in the relative interior of $\mathcal{LF}_\beta(\rho)$. Lemma 5.7 now implies all the Aubry sets $\mathcal{A}_{H_\epsilon}(\lambda c_1 + (1 - \lambda)c_2)$, $0 \leq \lambda \leq 1$ coincide, which implies (**Ma**) applies to the whole segment. The proposition follows.   □

5.4   JUMP FROM NON-SIMPLE HOMOLOGY TO SIMPLE HOMOLOGY

As described, when h is not a simple homology, the cohomology class $\Gamma^{DR}_{k_1,k_2}$ as chosen is not connected. More precisely, $\Gamma^{DR}_{k_1,k_2}$ consists of three connected components (see also (5.3) and (5.9)):

$$\Phi_L^* \bar{\Gamma}^e_h \cup \Gamma^{con}_+, \quad \Phi_L^* \bar{\Gamma}^e_{-h} \cup \Gamma^{con}_-, \quad \Phi_L^* \left(\bar{\Gamma}^{e,\mu}_{h_1} \cup \bar{\Gamma}^{e,\mu}_{-h_1} \right) \cup \Gamma^{con}_0.$$

We will show forcing equivalence of the components by the following:

Theorem 5.9 (Chapter 12). *Suppose the slow system H^s satisfies the condition $[DR1^c] - [DR4^c]$, and that the associated homology $h = n_1 h_1 + n_2 h_2$ is non-simple. Then there exist $e, \mu, \epsilon_0, \delta > 0$ depending on H^s, such that there is a residual subset $\mathcal{R}_\delta(H_1)$ of $\mathcal{V}_\delta(H_1)$, and for each $H_1' \in \mathcal{R}_\delta(H_1)$, there is $E_1, E_1 \in (e, e + \mu)$ such that*

$$\Phi_L^* \left(\bar{c}_h(E_1) \right), \quad \Phi_L^* \left(\bar{c}^{\mu}_{h_1}(E_2) \right)$$

are forcing equivalent, with respect to $H_0 + \epsilon H_1'$. The same conclusions apply when h, h_1 are replaced with $-h, -h_1$.

See the dashed line in Figure 5.2.

5.5   FORCING EQUIVALENCE AT THE DOUBLE RESONANCE

We summarize all of our constructions in the following theorem.

Theorem 5.10. *Suppose H_1 satisfies all non-degeneracy conditions of Γ_{k_1} and Γ_{k_1,k_2}. Then there is $\epsilon_1 = \epsilon_1(H_0, H_1) > 0$, $\delta = \delta(H_0, H_1) > 0$, such that the following hold for all $H_1' \in \mathcal{V}_\delta(H_1)$, and $0 < \epsilon < \epsilon_1$. There is a subset $\Gamma^{DR}_{k_1,k_2} = \Gamma^{DR}_{k_1,k_2}(H_0, H_1', \epsilon)$ satisfying*

- $\Gamma_{k_1,k_2} \subset \Gamma^{DR}_{k_1,k_2}$;
- $\Gamma^{DR}_{k_1,k_2}$ *intersects both $\Gamma^{SR}_{k_1,k_2,+}$ and $\Gamma^{SR}_{k_1,k_2,-}$,*

$\sigma = \sigma(H_0, H_1, \epsilon) > 0$, *and a residual subset $\mathcal{R}_\sigma(H_0 + \epsilon H_1')$ of $\mathcal{V}_\sigma(H_0 + \epsilon H_1')$, such that for all $H \in \mathcal{R}(H_0 + \epsilon H_1')$, all of $\Gamma^{DR}_{k_1,k_2}$ are forcing equivalent.*

Proof. The set $\Gamma^{DR}_{k_1,k_2}$ is defined in (5.9), and its two properties hold by construction. During the proof, we say a property holds "after a residual perturbation" if it holds on a residual subset of a neighborhood of the corresponding Hamiltonian. If finitely many properties each hold after a residual perturbation, then they hold simultaneously after a residual perturbation. Moreover, if a property holds after a residual perturbation, and is invariant under coordinate changes, then it also holds in the new coordinate after a residual perturbation, provided the coordinate change is smooth enough. This is the case for our system since all coordinate changes are C^∞.

Let δ be the smallest parameter depending on H^s such that all of Corollary 5.3, Proposition 5.5, and Proposition 5.8 hold.

Let $\delta_1 = \delta_1(H_0, H_1)$ be such that for any $H_1' \in \mathcal{V}_{\delta_1}(H_1)$, the associated slow system is in $\mathcal{V}_\delta(H^s)$. Then for $0 < \epsilon < \epsilon_0$, Lemma 14.7 implies the system $H_0 + \epsilon H_1'$ is reduced to a system $(G^s + \sqrt{\epsilon}P)/\beta$ satisfying the conclusions of Corollary 5.3. Moreover, by Proposition 5.4, relative to the normal form system N_ϵ, one of the three diffusion mechanisms holds for $\Phi_L^*(\bar{\Gamma}_h^{DR})$, after taking a residual perturbation. By Lemma 2.8, the same holds relative to the original system $H_0 + \epsilon H_1'$.

Recall that (see 5.9) Γ_{k_1,k_2}^{DR} consists of the set $\Phi_L^*(\bar{\Gamma}_h^{DR})$ and the connector sets Γ_0^{con} and $\Gamma_\pm^{\mathrm{con}}$. Propositions 5.5 and 5.8 implies forcing equivalence of the connectors after a residual perturbation. This implies each connected component of Γ_{k_1,k_2}^{DR} is in a single forcing equivalence class.

Finally, Theorem 5.9 implies all three components of Γ_{k_1,k_2}^{DR} are equivalent to each other after a residual perturbation. $\square$

Assuming all the propositions and theorems formulated thus far, we prove Theorem 2.2 which implies our main theorem.

Proof of Theorem 2.2. Let $H_1 \in \mathcal{U}$, which means that H_1 satisfies $SR(k_1, \lambda)$ for some $\lambda > 0$ and all $k_1 \in \mathcal{K}$, and that for all the strong double resonances $k_2 \in \bigcup_{k_1 \in \mathcal{K}} \bigcup \mathcal{K}^{\mathrm{st}}(k_1, \Gamma_{k_1}, \lambda)$, H_1 satisfies the non-degeneracy conditions $DR(k_1, \Gamma_{k_1}, k_2)$.

For each k_1, Theorem 3.3 applies. Therefore, there exists $\epsilon_1^{k_1}(H_0, \lambda) > 0$ such that the theorem applies for each $H_0 + \epsilon H_1$ with $0 < \epsilon < \epsilon_1$. Since $SR(k_1, \Gamma_{k_1}, \lambda)$ is an open condition, there exists $\delta^{k_1} = \delta^{k_1}(H_0, H_1)$ such that the conclusion of the theorem holds for all $H_1' \in \mathcal{V}_{\delta^{k_1}}(H_1)$.

For each k_1, k_2, the conclusion of Theorem 5.10 holds for all $H_1' \in \mathcal{V}_{\delta^{k_1,k_2}}(H_1)$ and $0 < \epsilon < \epsilon_1^{k_1,k_2}(H_0, H_1)$. Define

$$\epsilon_1(H_0, H_1) = \min\left\{ \min_{k_1,k_2} \epsilon_1^{k_1,k_2}, \min_{k_1} \epsilon_1^{k_1} \right\},$$

$$\delta(H_0, H_1) = \min\left\{ \min_{k_1,k_2} \delta^{k_1,k_2}, \min_{k_1} \delta^{k_1} \right\},$$

where the minimum is taken over all $k_1 \in \mathcal{K}$ and $k_2 \in \bigcup_{k_1 \in \mathcal{K}} \bigcup \mathcal{K}^{\mathrm{st}}(k_1, \Gamma_{k_1}, \lambda)$. ϵ_1 and δ are positive since all minima taken are over a finite set of positive numbers. The conclusions of Theorems 3.3 and 5.10 apply to $H_1' \in \mathcal{V}_\delta(H_1)$ and $0 < \epsilon < \epsilon_1$.

Define

$$\Gamma_*(H_0, H_1', \epsilon) = \bigcup_{k_1 \in \mathcal{K}} \left(\Gamma_{k_1}^{SR} \cup \bigcup_{k_2 \in \mathcal{K}^{st}(k_1, \lambda)} \Gamma_{k_1,k_2}^{DR} \right).$$

For each single resonance k_1, the union

$$\Gamma_*^{SR}(k_1) = \Gamma_{k_1}^{SR} \cup \bigcup_{k_2 \in \mathcal{K}^{st}(k_1,\lambda)} \Gamma_{k_1,k_2}^{DR}$$

is contained in a single equivalent class, since each Γ_{k_1,k_2}^{DR} is forcing equivalent, and they connect all the disconnected pieces from $\Gamma_{k_1}^{SR}$. If two single resonances Γ_{k_1} and $\Gamma_{k_1'}$ intersect at $\Gamma_{k_1,k_1'}$, then $\Gamma_*^{SR}(k_1)$ and $\Gamma_*^{SR}(k_1')$ also intersect at $\Gamma_{k_1,k_1'}$, since $\Gamma_{k_1,k_1'}$ is contained in both $\Gamma_{k_1,k_1'}^{DR}$ and Γ_{k_1',k_1}^{DR}. As a result, the entire $\Gamma_*(H_0, H_1', \epsilon)$ is contained in a single forcing equivalent class.

Finally, if $U_1, \ldots, U_N$ are open sets which intersect $\mathcal{P}$, by setting ϵ_0 small enough, they also intersect $\bigcup_{k_1 \in \mathcal{K}} \Gamma_{k_1}^{SR}$, since the said union is obtained from $\mathcal{P}$ by removing finitely many neighborhoods of size $O(\sqrt{\epsilon})$. Therefore, $U_1, \ldots, U_N$ also intersect $\Gamma_*(H_0, H_1', \epsilon)$. $\qquad\square$

Part II

Forcing relation and Aubry-Mather type

Chapter Six

Weak KAM theory and forcing equivalence

In this chapter we give an introduction to weak Kolmogorov-Arnold-Moser (KAM) theory and forcing relation. Most of the presentations follow [11] and [37]. One change from the standard presentation is that we need to modify the definition of Tonelli Hamiltonians to allow different periods in the t component.

6.1 PERIODIC TONELLI HAMILTONIANS

A C^2 Hamiltonian
$$H : \mathbb{T}^n \times \mathbb{R}^n \times \mathbb{R} \to \mathbb{R}$$
is called (time-periodic) Tonelli if it satisfies:

1. (Periodicity) There is $0 < \varpi = \varpi_H$ such that $H(\theta, p, t + \varpi) = H(\theta, p, t)$.
2. (Convexity) $\partial^2_{pp} H(x, p, t)$ is strictly positive definite as a quadratic form.
3. (Superlinearity) $\lim_{\|p\| \to \infty} H(x, p, t)/\|p\| = \infty$.
4. (Completeness) The Hamiltonian vector field generates a complete flow on $\mathbb{T}^n \times \mathbb{R}^n$. We denote by $\phi_H^{s,t}$ the flow from time s to time t, and by ϕ_H the flow $\phi_H^{0,\varpi}$.

The Lagrangian associated to H is
$$L = L_H(\theta, v, t) = \sup\{p \cdot v - H(\theta, p, t)\},$$
and the Hamiltonian flow is conjugate to the Euler-Lagrange flow via the Legendre transform
$$\mathbb{L} : (\theta, p, t) \mapsto (\theta, \partial_p H(\theta, p, t), t).$$

For the most part, we will restrict to Hamiltonians with $\varpi = 1$, namely, defined on $\mathbb{T}^n \times \mathbb{R}^n \times \mathbb{T}$, but near double resonances we need to consider Hamiltonians that are $\sqrt{\epsilon}$-periodic. It is helpful to consider a family of Hamiltonians which satisfy these properties uniformly. For $D > 0$, consider

$$\mathbb{H}(D) = \Big\{ H \in C^2(\mathbb{T}^n \times \mathbb{R}^n \times \mathbb{R}) :$$
$$\varpi_H \le 1, \quad D^{-1} I \le \partial^2_{pp} H(\theta, p, t) \le DI \quad \text{for all } p \in \mathbb{R}^n,$$
$$\|H(\cdot, 0, \cdot)\|_{C^0}, \|\partial_p H(\cdot, 0, \cdot)\|_{C^0} \le D \Big\}.$$

We then check that each $H \in \mathbb{H}(D)$ is Tonelli, and it satisfies a list of uniform estimates called *uniform family* in [11]. In particular, if $H_0 \in \mathbb{H}(D)$, then $H_\epsilon \in \mathbb{H}(2D)$ for all $\epsilon \le \epsilon_0 = \epsilon_0(D)$.

Given $C > 0$, and let $\Omega \subset \mathbb{R}^n$ be an open convex set. We say a function $u : \Omega \to \mathbb{R}$ is C semi-concave if for every $x \in \Omega$, there is a linear function $l_x : \mathbb{R}^n \to \mathbb{R}$, such that

$$u(y) - u(x) \le l_x(y - x) + C\|y - x\|^2, \quad y \in \Omega.$$

The linear form l_x is called a super-differential at x. The set of all super-differentials at x is denoted $\partial^+ u(x)$. It is easy to see that if u is differentiable at x, then $\partial^+ u(x) = \{du(x)\}$. A function $u : \mathbb{T}^n \to \mathbb{R}$ is semi-concave if it's semi-concave as a function on $\mathbb{R}^n$.

Lemma 6.1. *([11]) If $u : \mathbb{T}^n \to \mathbb{R}$ is C semi-concave, then it is $C\sqrt{n}$–Lipschitz. The super-differential set $\partial^+ u(x) \subset \{\|p\| \le C\sqrt{n}\}$.*

Given $s < t \in \mathbb{R}$, $x, y \in \mathbb{T}^n$, we define the Lagrangian action

$$A_H(x, s, y, t) = A_L(x, s, y, t) = \inf_{\gamma(s)=x,\, \gamma(t)=y} \int_s^t L_H(\gamma(\tau), \dot{\gamma}(\tau), \tau)d\tau, \quad (6.1)$$

where the infimum is taken over all absolutely continuous γ. We outline a series of useful results:

Proposition 6.2. *Let $H \in \mathbb{H}(D)$, then:*

1. *(Tonelli Theorem) (Appendix 2 of [60]) The infimum in (6.1) is always reached. It is then C^2, which solves the Euler-Lagrange equation. Such a γ is called a minimizer.*
2. *(A priori compactness) (Section B.2 of [11]) There is a constant $C_\delta > 0$ depending only on δ and D, such that for $t - s > \delta$, any minimizer of $A_H(x, s, y, t)$ satisfies $\|\dot{\gamma}\| \le C_\delta$.*
3. *(Uniform semi-concavity in space variable) (Theorem B.7 of [11]) For $t - s > \delta$, the function $A_H(x, s, y, t)$ is C_δ semi-concave in x and y. Moreover, if $\gamma : [s, t] \to \mathbb{T}^n$ is a minimizer, then*

$$p(s) \in -\partial_x^+ A_H(x, s, y, t), \quad p(t) \in \partial_y^+ A_H(x, s, y, t), \quad (6.2)$$

where $p(\tau) = \partial_v L_H(\gamma(\tau), \dot{\gamma}(\tau), \tau)$.
4. *(Uniform semi-concavity in the time variable, see Corollary 7.15) For $t - s > 2$, $A_H(x, s, y, \cdot)$ is C-semi-concave on $(t - 1, t + 1)$, and $A_H(x, \cdot, y, t)$ is C-semi-concave on $(s - 1, s + 1)$. If $\gamma : [s, t] \to \mathbb{T}^n$ is a minimizer,*

$$-H(\gamma(s), p(s), s) \in -\partial_s^+ A_H(x, s, y, t), \quad -H(\gamma(t), p(t), t) \in \partial_t^+ A_H(x, s, y, t). \quad (6.3)$$

In particular, by Lemma 6.1, we have $A_H(\cdot, \cdot, y, t)$ *and* $A_H(x, s, \cdot, \cdot)$ *are uniformly Lipschitz.*

Remark 6.3. Item 4 seems to be known but we have not been able to locate a reference. We give a proof in Corollary 7.15 along with some more refined results about semi-concavity of action functions.

For each $c \in \mathbb{R}^n$, define

$$L_{H,c}(\theta, v, t) = L_H(\theta, v, t) - c \cdot v,$$

and denote $A_{H,c} = A_{L_{H,c}}$. If $t - s > 2$, we have

$$(p(t) - c, -H(\gamma(t), p(t), t)) \in \partial^+_{(y,t)} A_{H,c}(x, s, y, t)$$

and similarly for s.

6.2   WEAK KAM SOLUTION

We now define the (continuous) Lax-Oleinik semi-group $T^{s,t}_c : C(\mathbb{T}^n) \to C(\mathbb{T}^n)$ via the formula

$$T^{s,t}_c u(x) = \min_{z \in \mathbb{T}^n} \{u(z) + A_{H,c}(z, s; x, t)\}.$$

The semi-group property $T^{t_1,t_2}_c T^{t_0,t_1}_c u = T^{t_0,t_2}_c u$ is clear from definition. For a ϖ-periodic Hamiltonian, the associated discrete semi-group is generated by the operator: $T_c u = T^{0,\varpi}_c u$.

Lemma 6.4. *(See (A.3) and (A.10) of [11]) Let $\{u_\zeta\}_{\zeta \in Z}$ be a (possibly uncountable) family of C-semi-concave functions $\mathbb{T}^n \to \mathbb{R}$, and $v = \inf u_\zeta$ is finite for at least one point, then $\inf_{\zeta \in Z} u_\zeta$ is also C semi-concave.*
   Moreover, suppose for $x_0 \in Z$, the infimum $v(x_0)$ is reached at $u_{\zeta_0}(x_0)$, then $\partial^+ u_{\zeta_0}(x_0) = \partial^+ v(x_0)$.

Using Proposition 6.2, the functions $T^n_\eta u$ for all $n > 1/\varpi$ are C semi-concave with the constant depending only on the uniform family $\mathbb{H}(D)$.

Proposition 6.5 (Mañé 's critical value, see Proposition 3.1 of [11]). *There is a unique $\alpha \in \mathbb{R}$ such that $T^n_c u(x) + n\alpha$ is uniformly bounded over all $n \in \mathbb{N}$. This value coincides with Mather's alpha function:*

$$\alpha_H(c) = -\inf_\mu \left\{ \int L_{H,c}(\theta, v, t) d\mu(\theta, v, t) \right\} \tag{6.4}$$

where the infimum is taken over all invariant probability measures of the Euler-Lagrange flow on $\mathbb{T}^n \times \mathbb{R}^n \times \mathbb{T}_\varpi$.

Let $\hat{H}(\theta, p, c) = H(\theta, p + c, t) - \alpha_H(c)$, then

$$L_{\hat{H}} = L_{H,c} + \alpha_H(c), \tag{6.5}$$

and $\alpha_{\hat{H}}(0) = 0$. This reduction is often used in proofs to reduce a statement about $L_{H,c}$ to the case $c = 0$ and $\alpha_H(0) = 0$.

Let us also point out an alternative definition of the alpha function, namely, we can replace the class of minimal measures with the class of *closed measures*, which satisfies

$$\int df(\theta, t) \cdot (v, 1) d\mu(\theta, v, t) = 0$$

for every C^1 function $f : \mathbb{T}^n \times \mathbb{R} \to \mathbb{R}$ satisfying $f(\theta, t + \varpi) = f(\theta, t)$.

Proposition 6.6 (See [74]).

$$\alpha_H(c) = -\inf_{\mu} \left\{ \int L_{H,c}(\theta, v, t) d\mu(\theta, v, t) \right\}$$

with μ taken over all closed probability measures.

A function $w : \mathbb{T}^n \times \mathbb{T}_{\varpi} \to \mathbb{R}$ is called a *weak KAM solution* if

$$T_c^{s,t} w(\cdot, s) + \alpha_H(c)(t - s) = w(\cdot, t), \quad s < t \in \mathbb{R},$$

i.e. the family $w(\cdot, t)$ is invariant under the semi-group up to a linear drift. The function $u(\theta) = w(\theta, 0)$ is then a fixed point of the operator $T_c + \varpi \alpha_H(c)$.

Proposition 6.7 (Existence of weak KAM solution, Proposition 3.2 of [11]). *For $c \in \mathbb{R}^n$ and $u_0 \in C(\mathbb{T}^n)$, the function*

$$w(\theta, t) = \liminf_{N \to -\infty} (T_c^{t - N\varpi, t} u_0 + N\varpi \alpha(c)), \quad N \in \mathbb{N}$$

is a ϖ-periodic weak KAM solution.

Lemma 6.8. *Weak KAM solutions are uniformly semi-concave over all $H \in \mathbb{H}(D)$.*

Proof. Let $w(x, t)$ be a weak KAM solution; we view it as a function on $\mathbb{T}^n \times \mathbb{R}$. The semi-concavity in x is clear by Lemma 6.4 and Proposition 6.2, item 3. We now prove the semi-concavity in t. By definition, for $t - s > 2$,

$$w(x, t) = \min_{z \in \mathbb{T}^n} \{w(z, s) + A_{H,c}(z, s, x, t) + \alpha_H(c)(t - s)\},$$

which by Proposition 6.2, item 4, is C-semi-concave on any interval $(t_0 - 1, t_0 + 1)$. Since $w(x, \cdot)$ has period $\varpi \in (0, 1]$, this implies it is also C-semi-concave on $\mathbb{R}$. $\qquad \square$

A function $w : \mathbb{T}^n \times \mathbb{R} \to \mathbb{R}$ is called *dominated* by $(L_{H,c}, \alpha)$ if

$$w(y, t) - w(x, s) \le A_{H,c}(x, s, y, t) + (t - s)\alpha, \quad x, y \in \mathbb{T}^n, \quad s < t.$$

$\gamma : I \to \mathbb{T}^n$, where I is an interval in $\mathbb{R}$, is called *calibrated* by w if

$$w(\gamma(t), t) - w(\gamma(s), s) = A_{H,c}(x, s, y, t) + (t - s)\alpha.$$

Proposition 6.9 (Proposition 4.1.8 and Theorem 4.2.6 of [37]). $w : \mathbb{T}^n \times \mathbb{R} \to \mathbb{R}$ *is a weak KAM solution if and only if it is dominated by* $(L_{H,c}, \alpha_H(c))$ *and for every* (y, t) *there is a calibrated curve* $\gamma : (-\infty, t] \to \mathbb{T}^n$ *such that* $\gamma(t) = y$.
    Moreover, for each $s \in (-\infty, t)$, $p(s) = \partial_v L_H(\gamma(s), \dot{\gamma}(s), s)$, $u(\cdot, s)$ *is differentiable at* $\gamma(s)$ *and* $\partial_x u(\gamma(s), s) = p(s) - c$. *The calibrated curve* γ *is unique if* $\partial_x u(\gamma(t), t)$ *exists.*

6.3   PSEUDOGRAPHS, AUBRY, MAÑÉ, AND MATHER SETS

Let $u : \mathbb{T}^n \to \mathbb{R}$ be semi-concave. By the Rademacher theorem, u is differentiable almost everywhere. For $c \in \mathbb{R}^n$, we define the (overlapping) pseudograph

$$\mathcal{G}_{c,u} = \mathcal{G}_{c,H,u} = \{(x, c + \nabla u(x)), \quad \nabla u(x) \text{ exists}\}.$$

In the 1-dimensional case, at every discontinuity of the function $du(x)$, the left limit is larger than the right limit. In the time-dependent setting if $w : \mathbb{T}^n \times \mathbb{R} \to \mathbb{R}$ is semi-concave, we write

$$\mathcal{G}_{c,w} = \{(x, c + \partial_x w(x, t), t) : \quad \partial_x w \text{ exists}\}.$$

The evolution by the Lax-Oleinik semi-group generates an evolution operator on the pseudograph. The following statement outlines its relation with the Hamiltonian dynamics.

Proposition 6.10 ([11]). *For each* $s < t$, *we have*

$$\overline{\mathcal{G}_{c, T_c^{s,t} u}} \subset \phi_H^{s,t}\left(\mathcal{G}_{c,u}\right),$$

here $\phi_H^{s,t}$ *denotes the Hamiltonian flow.*

Corollary 6.11. *Suppose* $w(\theta, t)$ *is a (time-periodic) weak KAM solution of* $L_{H,c}$, *then*

$$\left(\phi_H^{s,t}\right)^{-1} \overline{\mathcal{G}_{c, w(\cdot, t)}} \subset \mathcal{G}_{c, w(\cdot, s)}.$$

In particular, $(\phi_H^{-1}) \overline{\mathcal{G}_{c,u}} \subset \mathcal{G}_{c,u}$, *where* $u = w(\cdot, 0)$ *and* $\phi_H = \phi_H^{0, \varpi}$ *is the associated discrete dynamics.*

Let $w = w(\theta, t)$ be a continuous weak KAM solution for $L_{H,c}$. Then Corollary 6.11 implies the set

$$\mathcal{G}_{c,w} = \left\{(\theta, p) : \quad (\theta, p) \in \mathcal{G}_{c,w(\cdot,t)}\right\}$$

is backward invariant under the flow $\phi_H^{s,t}$. We then define

$$\widetilde{\mathcal{I}}(c, w) = \left\{(\theta, p, t) : \quad (\theta, p) \in \bigcap_{s<t} \left(\phi_H^{s,t}\right)^{-1} \overline{\mathcal{G}_{c,w(\cdot,t)}}\right\},$$

in other words, $\widetilde{\mathcal{I}}(c, w)$ is the invariant set generated by the family of pseudographs $\mathcal{G}_{c,w(\cdot,t)}$, in the extended phase space $\mathbb{T}^n \times \mathbb{R}^n \times \mathbb{T}_\varpi$.

The Aubry and Mañé sets admit the following equivalent definitions:

$$\widetilde{\mathcal{A}}(c) = \bigcap_w \widetilde{\mathcal{I}}(c, w), \quad \widetilde{\mathcal{N}}(c) = \bigcup_w \widetilde{\mathcal{I}}(c, w),$$

where $w : \mathbb{T}^n \times \mathbb{R} \to \mathbb{T}_\varpi$ is taken over all $L_{H,c}$ continuous-time weak KAM solutions. We now define the Mather set

$$\widetilde{\mathcal{M}}(c) = \bigcup\{\operatorname{supp}\mu : \operatorname{supp}\mu \subset \widetilde{\mathcal{A}}(c), \mu \text{ is invariant under } \phi_H^{s,t}\}.$$

Let's call u a discrete-time weak KAM solution if there is a (continuous-time) weak KAM solution $w : \mathbb{T}^n \times \mathbb{R} \to \mathbb{T}$ such that $u = w(\cdot, 0)$. Then analogous statements hold for $\widetilde{\mathcal{A}}^0$ and $\widetilde{\mathcal{N}}^0$ using discrete-time weak KAM solutions.

6.4 THE DUAL SETTING, FORWARD SOLUTIONS

There is a dual setting which corresponds to forward dynamics (as opposed to the backward invariant sets obtained before). Define

$$\check{T}_c^{s,t}u(x) = \max_z \{u(z) - A_c(x, s; z, t)\},$$

and note the following:

1. $-\check{T}_c^{s,t}u$ are uniformly semi-concave (if $t - s > \tau$). $w^+(x, t)$ is called a forward weak KAM solution if

$$\check{T}_c^{s,t}w(\cdot, t) = w(\cdot, s) - \alpha(t - s)$$

   for some $\alpha \in \mathbb{R}$ and all $s < t$.
2. The discrete-time version is $\check{T}_c = \check{T}_c^{0,\varpi}$.
3. $\alpha = \alpha_H(c)$ is the unique number such that a weak KAM solution may exist.
4. For a semi-concave function u, we define $\check{\mathcal{G}}_{c,u} = \{(\theta, c + \nabla_x u(\theta))\}$, and call it

an anti-overlapping pseudograph.

5. Analogs of the previous section apply with appropriate changes.

Let $w(\theta, t)$ be a weak KAM solution for $L_{H,c}$, and w^+ a forward weak KAM solution. We say that w, w^+ are conjugate if they coincide on the set $\mathcal{M}_H(c)$. Denote

$$\widetilde{\mathcal{I}}(c, w, w^+) := \{(x, c + \partial_x w(x, t), t) : \quad (x, t) \in \operatorname{argmin}(w - w^+)\}.$$

Proposition 6.12 (Theorem 5.1.2 of [37]). *For each weak KAM solution w, there exists a forward solution w^+ conjugate to w satisfying $w^+ \leq w$, and*

$$\widetilde{\mathcal{I}}(c, w) = \widetilde{\mathcal{I}}(c, w^+) = \widetilde{\mathcal{I}}(c, w, w^+).$$

One important consequence of Proposition 6.12 is that $\widetilde{\mathcal{I}}(c, w)$ is a Lipschitz graph over its x component. This is immediate given the next lemma.

Lemma 6.13 (Proposition 4.5.3 of [37]). *If $w, -w^+$ are C-semi-concave, then $\partial_x w$ is uniformly Lipschitz over the set*

$$\operatorname{argmin}_{(x,t)} \{w(x, t) - w^+(x, t)\}$$

where the Lipschitz constant depends only on C.

In fact, this approach gives a more general result which we will need later.

Proposition 6.14. *Let $w : \mathbb{T}^n \times \mathbb{T}_\varpi \to \mathbb{R}$ be a discrete-time weak KAM solution for H at cohomology c. For $t > s$, set*

$$w_k^+(x, s) = \check{T}_c^{s,t} w.$$

Then

$$\phi^{s-t} \overline{\mathcal{G}_{c,w}} \subset \widetilde{\mathcal{I}}(c, w, w_k^+).$$

As a consequence $\phi^{s-t} \overline{\mathcal{G}_{c,w}}$ are uniformly Lipschitz graphs over x component.

Proof. Suppose $(x_s, p_s, s) \in \phi^{s-t} \mathcal{G}_{c,w}$, and $(x_t, p_t, t) = \phi^{t-s}(x_s, p_s, s) \in \mathcal{G}_{c,w}$, then

$$w(x_t, t) = w(x_s, s) + A_{H,c}(x_s, s, x_t, t)$$

since the backward orbit of (x_t, p_t) is calibrated by Proposition 6.9. Then

$$(\check{T}_c^{s,t} w)(x_s) \geq w(x_s, s).$$

On the other hand, since w a weak KAM solution, for arbitrary x, y,

$$w(y, t) \leq w(x, s) + A_{H,c}(y, s, x, t).$$

This implies $\check{T}_c^{s,t} w(\cdot, t) \leq w(\cdot, s)$. It follows that $(x_s, s) \in \operatorname{argmin}(w - \check{T}^{s,t} w)$ and

$\phi^{s-t}\mathcal{G}_{c,w} \subset \widetilde{\mathcal{I}}(c, w, w_k^+)$. Since $\widetilde{\mathcal{I}}(c, w, w_k^+)$ is a compact set (since its a Lipschitz graph over a compact set $\mathrm{argmin}(w - T^{s,t}w))$, the inclusion also holds for the closure.   □

6.5   PEIERLS BARRIER, STATIC CLASSES, ELEMENTARY SOLUTIONS

We define

$$h_{H,c}(x, s, y, t) = \liminf_{N \to \infty} A_{H,c}(x, s, y, t + N\varpi) + N\varpi\alpha_H(c)$$

called the Peierls barrier by Mather [61]. Then $h_{H,c}$ projects to a function on $\mathbb{T}^n \times \mathbb{T}_\varpi \times \mathbb{T}^d \times \mathbb{T}_\varpi$. The functions $h_{H,c}(\cdot, \cdot, y, t)$ and $h_{H,c}(x, s, \cdot, \cdot)$ are uniformly Lipschitz, since they are limits of uniformly Lipschitz functions. We also have the triangle inequality:

$$h_{H,c}(x, s, y, t) + h_{H,c}(y, t, z, \tau) \geq h_{H,c}(x, s, z, \tau).$$

For any $(x, s) \in \mathbb{T}^n \times \mathbb{T}_\varpi$, we claim $h_{H,c}(x, s, x, s) \geq 0$. Suppose not, then we can always find a sequence of curves γ_k connecting (x, s) to itself such that the action goes to $-\infty$, contradicting Proposition 6.5. Then the projected Aubry set $\mathcal{A}_H(c)$ has the following alternative characterization:

$$\mathcal{A}_H(c) = \{(x, s) \in \mathbb{T}^n \times \mathbb{T}_\varpi : h_{H,c}(x, s, x, s) = 0\}.$$

We will also consider the discrete-time barrier:

$$h_{H,c}(x, y) = h_{H,c}(x, 0, y, 0), \tag{6.6}$$

then $\mathcal{A}_H^0(c) = \{x : h_{H,c}(x, x) = 0\}$.

Proposition 6.15. *For each $(x, s) \in \mathbb{T}^n \times \mathbb{T}_\varpi$, $h_{H,c}(x, s, \cdot, \cdot)$ is a weak KAM solution for $L_{H,c}$.*

Define the Mather semi-distance:

$$\bar{d}(x, s, y, t) = h_{H,c}(x, s, y, t) + h_{H,c}(y, t, x, s),$$

then the condition $\bar{d}(x, s, y, t) = 0$ defines an equivalence relation $(x, s) \sim (y, t)$ on $\mathcal{A}_H(c)$. The equivalence classes of this relation are called the *static classes*. Let $\mathcal{S} \subset \mathcal{A}_H(c)$ be a static class, it corresponds uniquely to an invariant set in the phase space:

$$\widetilde{\mathcal{S}} = \pi_{(x,t)}^{-1}\Big|_{\mathcal{A}_H(c)}\mathcal{S}.$$

For $(\zeta, \tau) \in \mathcal{S}$, the function

$$h_{H,c}(\zeta, \tau, \cdot, \cdot)$$

is independent of the choice of $(\zeta, \tau) \in \mathcal{S}$, up to an additive constant. These functions are called *elementary solutions*, due to the representation formula below.

Proposition 6.16.

1. *(Representation formula, see Theorem 8.6.1 of [37] and Theorem 7 of [29].) Let $w(x,t)$ be an $L_{H,c}$ weak KAM solution. Then*

$$w(x,t) = \min_{(\zeta,\tau) \in \mathcal{A}_H(c)} \{w(\zeta, \tau) + h_{H,c}(\zeta, \tau, x, t)\}.$$

2. *(See Proposition 4.3 of [11].) Every orbit in the Mañé set $\widetilde{\mathcal{N}}_H(c)$ is a heteroclinic orbit between two static classes $\widetilde{\mathcal{S}}_1, \widetilde{\mathcal{S}}_2$. We have $\widetilde{\mathcal{A}}_H(c) = \widetilde{\mathcal{N}}_H(c)$ if and only if $\widetilde{\mathcal{A}}_H(c)$ has only one static class.*

Suppose the Aubry set $\mathcal{A}_H(c)$ is the union of two non-empty compact components. The following lemma allows us to isolate each component by adding a bump function to the Hamiltonian.

Lemma 6.17. *Suppose $\mathcal{A}_1 \subset \mathcal{A}_H(c)$ and $\mathcal{A}_2 = \mathcal{A}_H(c) \setminus \mathcal{A}_1$ are both non-empty and compact, and $V \subset \mathcal{A}_1$ separates $\mathcal{A}_1$ and $\mathcal{A}_2$. Let $f \in C^\infty(\mathbb{T}^n \times \mathbb{T}_\varpi)$ be such that*

$$f|_V = 0, \quad \min f|_{\mathcal{A}_2} > 0.$$

Then for the Hamiltonian $H^f(x,p,t) = H(x,p,t) - f(\theta, t)$, we have

$$\alpha_H(c) = \alpha_{H^f}(c), \quad \mathcal{A}_{H^f}(c) = \mathcal{A}_1.$$

Proof. First we use (6.5) to reduce to the case $c = 0$ and $\alpha_H(0) = 0$. Given $(\theta_0, t_0) \in \mathcal{A}_1$, since $h_H(\theta_0, t_0, \theta_0, t_0) = 0$, there exists a sequence of curves $\gamma_k : I_k := [t_0, t_0 + n_k \varpi] \to \mathbb{T}^n$, such that $n_k \to \infty$, $\gamma(t_0) = \theta_0 = \gamma(t_0 + n_k \varpi)$, and

$$\lim_{k \to \infty} A_H(\gamma(t_0), t_0, \gamma(t_0 + n_k), t_0 + n_k) = 0.$$

We claim that, for all k sufficiently large, $\gamma_k(I_k) \subset V$. Suppose not, then there must exist $k_j \to \infty$ and $t_j \in T_{k_j}$ such that $(\gamma_k(t_j), t_j \mod \varpi \mathbb{Z}) \in \partial V$. By passing to a subsequence, we may assume $(\gamma_{k_j}(t_j), t_j) \to (y, s) \in \partial V$. Using the representation formula (Proposition 6.16),

$$h_H(\theta_0, t_0, y, s) \leq h_H(\theta_0, t_0, \theta_0, t_0) + A_H(\theta_0, t_0, \gamma_{k_j}(t_j + n_{k_j}), t_j).$$

Taking $\liminf$, we get $h_H(\theta_0, t_0, y, s) \leq 0$ and similarly $h_H(y, s, \theta_0, t_0) \leq 0$. It

follows that
$$0 \leq h_H(\theta_0, t_0, y, s) + h_H(y, s, \theta_0, t_0) \leq 0,$$
implying $(y, s) \in \mathcal{A}_H(0)$, a contradiction.

Using our claim, we now prove the lemma. Let L^f, L denote the Lagrangians for H^f and F, then $L^f = L + f$. First of all, since $L^f \geq L$, it follows from Proposition 6.6 that
$$\alpha_{H^f}(0) \leq \alpha_H(0) = 0.$$
We now have

$$0 \leq h_{H^f}(\theta_0, t_0, \theta_0, t_0) \leq \liminf_{k \to \infty} A_{L^f + \alpha_{H^f}(0)}(\theta_0, t_0, \gamma_k(t_0 + n_k), t_0 + n_k)$$
$$\leq \liminf_{k \to \infty} A_{L^f}(\theta_0, t_0, \gamma_k(t_0 + n_k), t_0 + n_k) \quad (\alpha_{H^f}(0) \leq 0)$$
$$= \liminf_{k \to \infty} A_L(\theta_0, t_0, \gamma_k(t_0 + n_k), t_0 + n_k) \quad (\text{since } \gamma_k(T_k) \subset V \text{ eventually})$$
$$\leq 0,$$

implying $(\theta_0, t_0) \in \mathcal{A}_{H^f}(0)$. The fact that $h_{H^f}(\theta_0, t_0, \theta_0, t_0)$ is bounded implies $\alpha_{H^f}(0) = \alpha_H(0) = 0$, concluding the proof. $\square$

6.6   THE FORCING RELATION

Definition 6.18. *Let $u : \mathbb{T}^n \to \mathbb{R}$ be semi-concave, and $N \in \mathbb{N}$. We say that $\mathcal{G}_{c,u} \vdash_N c'$, if there exists a semi-concave function $v : \mathbb{T}^n \to \mathbb{R}$ such that*

$$\overline{\mathcal{G}_{c',v}} \subset \bigcup_{k=0}^{N} \phi_H^k\left(\mathcal{G}_{c,u}\right).$$

We say the $c \vdash c'$ if there exists $N \in \mathbb{N}$ such that

$$\mathcal{G}_{c,u} \vdash_N c'$$

for every pseudograph $\mathcal{G}_{c,u}$. We say $c \dashv\vdash c'$ if $c \vdash c'$ and $c' \vdash c$.

In view of Proposition 6.10, we always have $c \vdash c$. The relation $\vdash$ is transitive by definition (but not symmetric). The relation $\dashv\vdash$ is an equivalence relation.

We summarize the property of the forcing relation in the text that follows.

- (Proposition 2.1)
  Let $\{c_i\}_{i=1}^N$ be a sequence of cohomology classes which are forcing equivalent. For each i, let U_i be neighborhoods of the discrete-time Mather sets $\widetilde{\mathcal{M}}_H^0(c_i)$; then there is a trajectory of the Hamiltonian flow ϕ^t of H visiting all the sets U_i.
- (Mather mechanism, Proposition 2.3)

Suppose $\mathcal{N}_H^0(c)$ is contractible as a subset of $\mathbb{T}^2$, then there is $\sigma > 0$ such that c is forcing equivalent to all $c' \in B_\sigma(c)$.

- (Arnold and bifurcation mechanism, Proposition 2.5)
  Suppose that either $\widetilde{\mathcal{A}}_H^0(c)$ has only two static classes and $\widetilde{\mathcal{N}}_H^0(c) \setminus \widetilde{\mathcal{A}}_H^0(c)$ is totally disconnected, or $\widetilde{\mathcal{A}}_H^0(c)$ has only one static class, and there is a symplectic double covering map ξ such that $\widetilde{\mathcal{N}}_{H\circ\Xi}^0(\xi^*c) \setminus \Xi^{-1}\widetilde{\mathcal{N}}_H^0$ is totally disconnected, then there is $\sigma > 0$ such that c is forcing equivalent to all $c' \in B_\sigma(c)$.

6.7 THE GREEN BUNDLES

In many parts of the proof, we study the hyperbolic property of a minimizing orbit, for which the concept of *Green bundles* is very useful. In this setting, we restrict to *autonomous* Tonelli Hamiltonians H on $\mathbb{T}^n \times \mathbb{R}^n$. Consider $\mathbb{V} = \{0\} \times \mathbb{R}^n \subset \mathbb{R}^n \times \mathbb{R}^n$, viewed as a sub-bundle of $T(\mathbb{T}^n \times \mathbb{R}^n)$, called the *vertical bundle*. An orbit $(\theta, p)(t)$ of the Hamiltonian flow $\phi_H^{s,t}$ is called *disconjugate* if

$$D(\phi_H^{s,t})\mathbb{V} \pitchfork \mathbb{V}, \quad \forall s \neq t \in \mathbb{R}.$$

Proposition 6.19 (See for example [3]). *Let $(\theta, p)(t)$, $t \in \mathbb{R}$ be a minimizing orbit for L_H. Then $(\theta, p)(t)$ is disconjugate, and the following hold:*

1. *For every t, the limits*

$$\mathcal{G}_-(\theta(t), p(t)) = \lim_{s \to -\infty} D(\phi_H^{s,t})(\theta(s), p(s))\mathbb{V}$$

$$\mathcal{G}_+(\theta(t), p(t)) = \lim_{s \to \infty} D(\phi_H^{s,t})(\theta(s), p(s))\mathbb{V}$$

exist. The bundles $\mathcal{G}_\pm$, called the Green bundles, are given by graphs of symmetric matrices $G_\pm$ with uniformly bounded norms depending only on H.

2. *If $(\theta, p)(t)$ is a hyperbolic orbit, then $\mathcal{G}_-$ coincide with the span of the unstable bundle and the flow direction, and $\mathcal{G}_+$ coincide with the span of the stable bundle and the flow direction.*

Chapter Seven

Perturbative weak KAM theory

By perturbative weak KAM theory, we mean two things:

- How do the weak KAM solutions and the Mather, Aubry and Mañé sets respond to limits of the Hamiltonian?
- How do the weak KAM solutions change when we perturb a system, in particular, what happens when we perturb (1) completely integrable systems, and (2) autonomous systems by a time-periodic perturbation?

In this chapter, we state and prove results in both aspects, as a technical tool for proving forcing equivalence.

7.1   SEMI-CONTINUITY

Let $\mathbb{H}(D)$ be the uniform family defined before, note that they are periodic of period $0 < \varpi_H \leq 1$, but not necessarily of the same period. It is known that in the case that all periods are fixed at 1, the weak KAM solutions are upper semi-continuous (see precise statements in the text that follows) under C^2 convergence over compact sets ([12]). The results generalize to the case when the periods are not the same, as we now show.

Let us remark that if $H_n \in \mathbb{H}(D)$ is a family of Hamiltonians, and $H_n \to H$ uniformly over compact sets on $\mathbb{T}^n \times \mathbb{R}^n \times \mathbb{R}$, then H is necessarily periodic of *some* period, and therefore $H \in \mathbb{H}(D)$.

Let A_k be a sequence of sets, define $\limsup_{k \to \infty} A_k$ to be the set of all accumulation points of all sequences $x_k \in A_k$.

Lemma 7.1 (Lemma 7 of [12]). *Suppose Hamiltonians $H_k \in \mathbb{H}(D)$, $k \in \mathbb{N}$ are a family of periodic Tonelli Hamiltonians. Suppose $H_n \to H$ in C^2 over compact sets, $c_k \to c \in \mathbb{R}^n$, then $\alpha_{H_k}(c_k) \to \alpha_H(c)$. If $w_k : \mathbb{T}^n \times \mathbb{R} \to \mathbb{R}$ is a sequence of weak KAM solutions of H_k, c_k which converges uniformly to $w : \mathbb{R}^n \times \mathbb{R} \to \mathbb{R}$, then w is a weak KAM solutions of H, c.*

Moreover, we have

$$\limsup_{k \to \infty} \mathcal{G}_{c_k, H_k, w_k} \subset \mathcal{G}_{c, H, w}, \quad \limsup_{k \to \infty} \tilde{\mathcal{I}}_{H_k}(c_k, w_k) \subset \tilde{\mathcal{I}}_H(c, w),$$

$$\limsup_{k \to \infty} \tilde{\mathcal{N}}_{H_k}(c_k) \subset \tilde{\mathcal{N}}_H(c).$$

Proof. The proof is an elaboration of Lemma 7 of [12]. Since all w_k are defined up to an additive constant, we assume in addition that $w_k(0,0) = w(0,0) = 0$.

Note that if $H_k \in \mathbb{H}$ and c_k uniformly bounded, then $\alpha_{H_k}(c_k)$ is uniformly bounded (6.4). By restricting to a subsequence, we may assume $\alpha_{H_k}(c_k) \to \alpha \in \mathbb{R}$. Then by taking the limit in

$$w_k(\gamma(t),t) - w_k(\gamma(s),s) \leq \int_s^t L_{H_k}(\gamma(\tau), \dot{\gamma}(\tau), \tau) - c_k \cdot \dot{\gamma}(\tau) + \alpha_{H_k}(c_k) \, d\tau,$$

we obtain

$$w(\gamma(t),t) - w(\gamma(s),s) \leq \int_s^t L_H(\gamma(\tau), \dot{\gamma}(\tau), \tau) - c \cdot \dot{\gamma}(\tau) + \alpha \, d\tau.$$

Using boundedness of w, we obtain

$$-\lim_{t-s \to \infty} \frac{1}{t-s} \int_s^t L_H(\gamma(\tau), \dot{\gamma}(\tau), \tau) - c \cdot \dot{\gamma}(\tau) \, d\tau \leq \alpha \tag{7.1}$$

for any γ defined on $\mathbb{R}$. If μ is any ergodic invariant measure for the Euler-Lagrange flow, for μ almost every orbit γ, the left-hand side of (7.1) converges to $-\int(L-c)d\mu$. We get $\alpha \geq \alpha_H(c)$ in view of (6.4).

Now given $x \in \mathbb{T}^n$, $t \in \mathbb{R}$, let $\gamma_k : (-\infty, t] \to \mathbb{T}^d$ be a L_{H_k,c_k}, w_k calibrated curve, then Proposition 6.2 implies γ_k are uniformly Lipschitz. Then γ_k has a subsequence that converges in C^1_{loc} to a limit $\gamma(t)$, which is $L_{H,c} + \alpha$, u calibrated. This implies both $\alpha_H(c) = \alpha$ and that u is a H,c weak KAM solution.

We now prove the "moreover" part. Denote

$$\mathcal{G}_k = \mathcal{G}_{c_k,H_k,w_k}, \quad \mathcal{G} = \mathcal{G}_{c,H,w} \quad \tilde{\mathcal{I}}_k = \tilde{\mathcal{I}}_H(c_k w_k), \quad \mathcal{I} = \mathcal{I}_H(c,w).$$

then if $(x_k, p_k, t_k) \in \mathcal{G}_k$, there exist L_{H_k,c_k}, w_k calibrated curves $\gamma_k : (-\infty, t_k] \to \mathbb{T}^d$ with $\gamma_k(t_k) = x_k$ and $\partial_p H_k(x_k, p_k, t_k) = \dot{\gamma}_n$. Then by the same argument, after restricting to a subsequence, γ_n converges in C^1_{loc} to $\gamma : (-\infty, t] \to \mathbb{T}^d$, and γ is a $L_{H,c}$, w calibrated curve. This implies $(t,x,p) \in \mathcal{G}$.

For the set $\mathcal{I}$, let us prove that for each fixed T, we have

$$\limsup_{k \to \infty} \phi_{H_k}^{-T} \mathcal{G}_k \subset \phi_H^{-T} \mathcal{G}.$$

Indeed, for any $(x_k, p_k, t_k) \in \phi_{H_k}^{-T} \mathcal{G}_{c_k,H_k,w_k}$, there exists L_{H_k,c_k}, w_k calibrated curves $\gamma : (-\infty, t_k + T] \to \mathbb{T}^d$ such that $\gamma(t_k) = x_k$, $\partial_p H_k(x_k, p_k, t_k) = \dot{\gamma}_k(t_k)$. Then exactly the same argument as before implies γ_n accumulates to a $L_{H,c}$, u calibrated curve $\gamma : (-\infty, t + T) \to \mathbb{T}^d$, and implying $(t,x,p) \in \Phi_H^{-T} \mathcal{G}$. We obtain

$$\limsup_{k\to\infty} \tilde{\mathcal{I}}_k \subset \tilde{\mathcal{I}}.$$

Finally, we have

$$\tilde{\mathcal{N}}_{H_k}(c_k) = \bigcup_w \tilde{\mathcal{I}}_H(c_k, w),$$

where the union is over all L_{H_k,c_k} weak KAM solutions w. For $(x_k, p_k, t_k) \in \tilde{\mathcal{N}}_{H_k}(c_k)$, there exist u_k such that $(x_k, p_k, t_k) \in \mathcal{G}_{c_k, H_k, u_k}$. All w_k are semi-concave with respect to a uniform constant, and hence they are equi-Lipschitz. Since all are normalized to $w_k(0, 0) = 0$, periodicity implies that they are equi-bounded. Ascoli's theorem implies there exists a subsequence that converges to u uniformly on the interval $\mathbb{T}^n \times [0, K]$. Since all w_k's are periodic with period bounded by 1, this implies $w_k \to w$ on $\mathbb{T}^n \times \mathbb{R}$ as well. We then apply the semi-continuity of $\mathcal{G}$ sets to get semi-continuity of $\tilde{\mathcal{N}}$. □

The theory built in [12] allows one to pass from semi-continuity of pseudo-graphs to semi-continuity of the Aubry set under a condition called the *coincidence hypothesis*. A sufficient condition for this hypothesis is when the Aubry set has finitely many static classes.

Corollary 7.2. *Suppose $\tilde{\mathcal{A}}_H(c)$ has at most finitely many static classes. Then if $H_k \in \mathbb{H}(D)$ C^2 converges to H over compact sets, and $c_k \to c$, we have*

$$\limsup_{n\to\infty} \tilde{\mathcal{A}}_{H_k}(c_k) \subset \tilde{\mathcal{A}}_H(c)$$

as subsets of $\mathbb{T}^n \times \mathbb{R}^n \times \mathbb{R}$.

Proof. The proof follows in the same way as the proof of Theorem 1 in [12]. We also refer to Section 6.2 of [52] where this is carried out in detail. □

7.2 CONTINUITY OF THE BARRIER FUNCTION

In general, the barrier function $h_{H,c}$ may be discontinuous with respect to H and c. However, the continuity properties hold in the particular case when the limiting Aubry set contains only one static class.

Proposition 7.3. *Assume that a sequence $H_k \in \mathbb{H}(D)$ converges to H in C^2 over compact sets, and $c_k \to c \in \mathbb{R}^n$. Assume that the projected Aubry set $\mathcal{A}_H(c)$ contains a unique static class. Let $(x_k, s_k) \in \mathcal{A}_{H_k}(c_k)$ with $(x_k, s_k) \to (x, s)$, then the barrier functions $h_{H_k, c_k}(x_k, s_k; \cdot, \cdot)$ converges to $h_{H,c}(x, s; \cdot, \cdot)$ uniformly.*

Similarly, for $(y_k, t_k) \in \mathcal{A}_{H_k}(c_k)$ and $(y_k, t_k) \to (y, t)$, the barrier functions $h_{H_k, c_k}(\cdot, \cdot; y_k, t_k)$ converge to $h_{H,c}(\cdot, \cdot; y, t)$ uniformly.

Proposition 7.4. *Assume that a sequence H_n converges to H in C^2 over compact sets, $c_k \to c \in \mathbb{R}^n$ and the Aubry set $\mathcal{A}_H(c)$ contains a unique static class.*

1. *For any $(x,s) \in \mathcal{A}_H(c)$ we have*

$$\lim_{k\to\infty} \sup_{(y,t)\in M\times\mathbb{T}} |h_{H_k,c_k}(x_k,s_k;y,t) - h_{H,c}(x,s;y,t)| = 0$$

uniformly over $(x_k,s_k) \in \mathcal{A}_{H_k}(c_k)$ and $(y,t) \in M \times \mathbb{R}$.
2. *For any $l_k \in \partial_y^+ h_{H_k,c_k}(x_k,s_k;y,t)$ and $l_k \to l$, we have $l \in \partial_y^+ h_{H,c}(x,s;y,t)$. Moreover, the convergence is uniform in the sense that*

$$\lim_{k\to\infty} \inf_{l_k\in\partial_y^+ h_{H_k,c_k}(x_k,s_k;y,t)} d(l_k, \partial_y^+ h_{H,c}(x,s;y,t)) = 0$$

uniformly in $(x,s) \in \mathcal{A}_H(c)$, $(x_k,s_k) \in \mathcal{A}_{H_k}(c_k)$.

The following statement follows easily from the representation formula (see Proposition 6.16).

Lemma 7.5. *Assume that $\mathcal{A}_H(c)$ has a unique static class. Let $(x,t) \in \mathcal{A}_H(c)$, then any weak KAM solution differs from $h_{H,c}(x,t;\cdot,\cdot)$ by a constant.*

Proof of Proposition 7.3. We prove the second statement. By Proposition 6.2, all functions $h_{H_k,c_k}(x_k,s_k;\cdot,\cdot)$ are uniformly semi-concave and equi-Lipschitz. Since $(x_k,s_k) \in \mathcal{A}_H(c_k)$ implies $h_{H_k,c_k}(x_k,s_k;x_k,s_k) = 0$, the functions are equi-bounded. By Ascoli's Theorem, the sequence $h_{H_k,c_k}(x_k,s_k;\cdot,\cdot)$ is pre-compact in uniform topology, and $h_{H,c}(x,s,\cdot,\cdot)$ is the unique accumulation point of the sequence. Indeed, due to Lemmas 7.1 and 7.5 any accumulation point must be

$$h_{H,c}(x,s;\cdot,\cdot) + C$$

for some $C \in \mathbb{R}$. But

$$h_{H,c}(x_k,s_k;x,s) \to h_{H,c}(x,s;x,s) = 0,$$

so $C = 0$.

Statement 1 follows from the definition of the projected Aubry set

$$\mathcal{A}_H(c) = \{(x,s) \in M \times \mathbb{T} : h_{H,c}(x,s;x,s) = 0\}$$

and statement 2. $\qquad\square$

Proof of Proposition 7.4. Part 1. We argue by contradiction. Assume that there exist $\delta > 0$, and by restricting to a subsequence,

$$\sup_{(y,t)} |h_{H_k,c_k}(x_k,s_k,y,t) - h_{H,c}(x,s,y,t)| > \delta.$$

By compactness, and by possibly restricting to a subsequence again, we may assume that $(x_k, s_k) \to (x^*, s^*)$, $(y_k, t_k) \to (y, t)$.

Using Proposition 7.3, take the limit as $n \to \infty$, we have

$$\sup_{(y,t)} |h_{H,c}(x^*, s^*, y, t) - h_{H,c}(x, s, y, t)| > \delta.$$

By Lemma 7.5, the left-hand side is 0, which is a contradiction.

Part 2. $h_{H_k, c_k}(x_k, s_k, \cdot, t)$ converges to $h_{H,c}(x, s, \cdot, t)$ uniformly. Convergence of super-differentials follows directly from Proposition 6.2. It suffices to prove uniformity.

Assume, by contradiction, that by restricting to a subsequence, we have $(x_k, s_k) \to (x, s) \in \mathcal{A}_H(c)$, $l_k \in \partial_y^+ h_{H_k, c_k}(x_k, s_k, y, t)$ and $(x, s) \in \mathcal{A}_H(c)$ such that

$$\lim_{k \to \infty} l_k \notin \partial_y^+ h_{H,c}(x, s, y, t).$$

By Proposition 6.2, $l_n \to l \in \partial_y^+ h_{H,c}(x^*, s^*, y, t)$. Moreover, $\partial_y^+ h_{H,c}(x, s, y, t) = \partial_y^+ h_{H,c}(x^*, s^*, y, t)$. Indeed, both functions $h_{H,c}(x, s, \cdot, \cdot)$ and $h_{H,c}(x_*, s_*, \cdot, \cdot)$ are weak KAM solutions, Lemma 7.5 implies they differ by a constant. This is a contradiction. □

7.3   LIPSCHITZ ESTIMATES FOR NEARLY INTEGRABLE SYSTEMS

In this section we record uniform estimates for the nearly integrable system

$$H_\epsilon = H_0(p) + \epsilon H_1(\theta, p, t), \quad (\theta, p, t) \in \mathbb{T}^n \times \mathbb{R}^n \times \mathbb{T},$$

with the assumptions $H_0 \in \mathbb{H}(D)$, $\|H_1\|_{C^2} \leq 1$.

Proposition 7.6 (Proposition 4.3, [13])**.** *For H_ϵ as given, for any $c \in \mathbb{R}^n$, any $L_{H_\epsilon, c}$ weak KAM solution $u(x, t)$ is $6D\sqrt{\epsilon}$-semi-concave and $6D\sqrt{n\epsilon}$-Lipschitz in x.*

Corollary 7.7. *For any $L_{H_\epsilon, c}$ weak KAM solution $u : \mathbb{T}^n \times \mathbb{T} \to \mathbb{R}$, let $\gamma : (-\infty, t_0] \to \mathbb{T}^n$ be a calibrated curve. Then for any $t \in (-\infty, t_0)$,*

$$p(t) := \partial_v L_{H_\epsilon}(\gamma(t), \dot{\gamma}(t), t)$$

satisfies $\|p(t) - c\| \leq 6D\sqrt{n\epsilon}$.

Moreover, any $(x, p, t) \in \widetilde{\mathcal{N}}_{H_\epsilon}(c)$ satisfies the same estimate $\|p - c\| \leq 6D\sqrt{n\epsilon}$.

Proof. Proposition 6.9 implies $p(t) - c = \partial_x u(\gamma(t), t)$, the conclusion follows from the fact that $u(\cdot, t)$ is $6D\sqrt{n\epsilon}$-Lipschitz. The same holds for $p(t_0)$ by continuity.

The statement for $\widetilde{\mathcal{N}}$ follows from the fact that any $(x, p, t) \in \mathcal{G} \subset \mathcal{N}$ is the end point of a calibrated curve. $\qquad\square$

Corollary 7.8. *Let μ be any c-minimal measure of H_ϵ, then there is $C > 0$ depending only on H_0 such that*

$$\|\rho(\mu) - \nabla H_0(c)\| \leq C\sqrt{\epsilon}.$$

Proof. Recall that μ is a measure on $\mathbb{T}^n \times \mathbb{R}^n \times \mathbb{R}$, invariant under the Euler-Lagrange flow. By considering the Legendre transform $\mathbb{L}$, we obtain

$$\rho(\mu) = \int v d\mu(\theta, v, t) = \int \partial_p H_\epsilon(\theta, p, t) d\mathbb{L}_* \mu(\theta, p, t).$$

By Corollary 7.7, we have $\partial_p H_\epsilon(\theta, p, t) = \nabla H_0(c) + O(\sqrt{\epsilon})$ for all

$$(\theta, p, t) \in \operatorname{supp} \mathbb{L}_* \mu \subset \widetilde{\mathcal{A}}_{H_\epsilon}(c),$$

therefore, $\rho(\mu) = \nabla H_0(c) + O(\sqrt{\epsilon})$. $\qquad\square$

7.4 ESTIMATES FOR NEARLY AUTONOMOUS SYSTEMS

The goal of this section is to derive a special Lipshitz estimate of weak KAM solutions for perturbations of autonomous systems. More precisely, consider

$$H_\epsilon(x, p, t) = H_1(x, p) + \epsilon H_2(x, p, t).$$

Assume that $H_1 \in \mathbb{H}(D/2)$ and $\|H_2\|_{C^2} = 1$. Then for ϵ small enough all $H_\epsilon \in \mathbb{H}(D)$. Let L_ϵ denote the associated Lagrangian.

We first state an estimate for the alpha function.

Lemma 7.9. *There is $C > 0$ depending only on $\|H_1\|_{C^2}$ and D, such that*

$$\|\alpha_{H_\epsilon}(c) - \alpha_{H_1}(c)\| < C\epsilon.$$

Proof. Let C always denote a generic constant depending on $\|H_1\|_{C^2}$. Let L_1 be the Lagrangian for H_1; then $\|L_\epsilon - L_1\|_{C^0} \leq C\epsilon$. As a result, the functionals

$$\int (L_\epsilon - c \cdot v) d\mu, \quad \int (L - c \cdot v) d\mu,$$

defined on the space of closed probability measures, differ by at most $C\epsilon$. The lemma follows immediately using the closed measure version for the definition of the alpha function, see Section 6.2. $\qquad\square$

Theorem 7.10. *Let $w : \mathbb{T}^n \times \mathbb{T}_{\varpi} \to \mathbb{R}$ be a weak KAM solution of $L_\epsilon - c \cdot v$ and suppose $t > 1/\sqrt{\epsilon}$. Then there is C depending only on D and $\|c\|$ such that*

$$|H_\epsilon(x_1, p_1, t_1) - H_\epsilon(x_2, p_2, t_2)| \leq C\sqrt{\epsilon}\,\|(x_2 - x_1, t_2 - t_1)\|, \quad (x_i, p_i, t_i) \in \phi^{-t}\mathcal{G}_{c,w}.$$

In particular, the above estimate holds on the Aubry sets $\widetilde{\mathcal{A}}_{L_\epsilon}(c)$.

The proof relies on semi-concavity of weak KAM solution, following Fathi ([37]). However to get an improved estimate we need a notion of semi-concavity that is "stronger" in the t direction.

Let $\Omega \subset \mathbb{R}^n$ be an open convex set. Let A be a symmetric $n \times n$ matrix. We say that $f : \Omega \to \mathbb{R}$ is A-semi-concave if for each $x \in \mathbb{R}^n$, there is $l_x \in \mathbb{R}^n$ such that

$$f(y) - f(x) - l_x \cdot (y - x) \leq \frac{1}{2}A(y - x)^2, \quad x, y \in \Omega$$

where Ax^2 denotes $Ax \cdot x$. This definition generalizes the standard semi-concavity, as A-semi-concave functions are $\frac{1}{2}\|A\|$-semi-concave. We say f is A-semi-convex if $-f$ is A-semi-concave. The following lemma follows from a direct computation.

Lemma 7.11. *f is A-semi-concave if and only if $f_A(x) = f(x) - \frac{1}{2}Ax^2$ is concave.*

The following lemma is proved in [80].

Lemma 7.12 (See Lemma 3.2 of [80]). *Suppose $f : \mathbb{R}^n \to \mathbb{R}$ is B-semi-concave and $g : \mathbb{R}^n \to \mathbb{R}$ is $(-A)$-semi-convex, and $S = B - A$ is positive definite. Suppose $f(x) \geq g(x)$ and M is the set on which $f - g$ reaches its minimum. Then for all $x_1, x_2 \in M$, we have*

$$\left\|df(x_2) - df(x_1) - \frac{1}{2}(A + B)(x_2 - x_1)\right\|_{S^{-1}} \leq \frac{1}{2}\|x_2 - x_1\|_S,$$

where $\|x\|_S = \sqrt{Sx^2}$.

Recall that the action function $A_L(x, s, y, t)$ is the minimal Lagrangian action of curves with end points $\gamma(s) = x$, $\gamma(t) = y$.

Proposition 7.13. *For the Hamiltonian H_ϵ and $c \in \mathbb{R}^n$, there is a constant $C > 0$ depending only on D and $\|c\|$ such that if $t_1 - t_0 > 1/\sqrt{\epsilon}$, for all $x_0, y_0 \in \mathbb{T}^n$, the action function $A_{H_\epsilon, c}(x_0, t_0, \cdot, \cdot)$ (treated as a function on $\mathbb{R}^n \times \mathbb{R}$) is S_ϵ-semi-concave on the set $(y, t) \in \mathbb{R}^n \times (t_1 - 1/(2\sqrt{\epsilon}), t_1 + 1/(2\sqrt{\epsilon}))$, and $A_{H_\epsilon, c}(\cdot, \cdot, y_0, t_1)$ on the set $(x, s) \in \mathbb{R}^n \times (t_0 - 1/(2\sqrt{\epsilon}), t_0 + 1/(2\sqrt{\epsilon}))$, where*

$$S_\epsilon = C \begin{bmatrix} \mathrm{Id}_{n \times n} & 0 \\ 0 & \sqrt{\epsilon} \end{bmatrix}.$$

We need the following lemma.

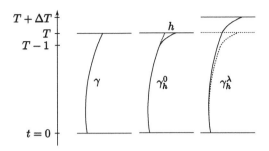

Figure 7.1: Defining the curve γ_h^λ.

Lemma 7.14. *Suppose $H \in \mathbb{H}(D)$, and let A be the action function defined using the Lagrangian L associated to H. Then there exists a constant C depending on D such that if $\gamma : [t_0, t_1] \to \mathbb{T}^n$ is an extremal curve with $T = t_1 - t_0 > 1$, then for $h \in \mathbb{R}^n$, $\Delta T \in [-T/2, T/2]$, $p(t) = \partial_v L(\gamma(t), \dot\gamma(t), t)$, $\kappa = \|\partial^2_{(x,v,t)t} L\|_{C^0}$,*

$$A(\gamma(t_0), t_0, \gamma(t_1) + h, t_1 + \Delta T) - A(\gamma(t_0), t_0, \gamma(t_1), t_1)$$

$$\leq p(t_1) \cdot h - H(\gamma(t_1), p(t_1), t_1) \cdot (\Delta T) + C\|h\|^2 + C\left(\frac{1}{T} + \kappa T\right)(\Delta T)^2.$$

Proof. By considering the Lagrangian $L(\cdot, \cdot, \cdot + t)$, it suffices to consider $t = 0$, $s = T$.

Let $\gamma : [0, T] \to \mathbb{T}^n$ be an absolutely continuous curve, write $\lambda = \Delta T/T$ and define

$$\gamma_h^\lambda : [0, T + \Delta T] \to \mathbb{T}^n,$$

$$\gamma_h^\lambda(t) = \begin{cases} \gamma\left(\frac{t}{1+\lambda}\right), & t \in [0, (T-1)(1+\lambda)]; \\ \gamma\left(\frac{t}{1+\lambda}\right) + \left(\frac{t}{1+\lambda} - T + 1\right)h, & t \in [(T-1)(1+\lambda), T(1+\lambda)]. \end{cases}$$

The curve is obtained by adding a linear drift in x on the time interval $[T-1, T]$, then reparametrize time to the interval $[0, T + \Delta T]$, see Figure 7.1.

Let C denote an unspecified constant. Note that $\|\dot\gamma\| \leq C$.

$$\mathbb{A}_L(\gamma_h^\lambda) = \int_0^{(1+\lambda)T} L(\gamma_h^\lambda, \dot\gamma_h^\lambda, t) dt$$

$$= (1+\lambda) \int_0^{T-1} L\left(\gamma(s), \frac{1}{1+\lambda}\dot\gamma(s), (1+\lambda)s\right) ds$$

$$+ (1+\lambda) \int_{T-1}^T L\left(\gamma(s) + (s - T + 1)h, \frac{1}{1+\lambda}(\dot\gamma(s) + h), (1+\lambda)s\right) ds;$$

then

$$\mathbb{A}_L(\gamma_h^\lambda)$$

$$\leq (1+\lambda) \int_0^T L(\gamma, \dot\gamma, s) ds + \int_0^{T-1} \left(-\lambda \partial_v L \cdot \dot\gamma + (1+\lambda) \partial_t L \cdot \lambda s \right) ds$$

$$+ (1+\lambda) \int_{T-1}^T \left(\partial_x L \cdot (s-T+1)h + \frac{1}{1+\lambda} \partial_v L \cdot (-\lambda \dot\gamma + h) + \partial_t L \cdot \lambda s \right) ds$$

$$+ CT \left(\|\partial_{vv}^2 L\| + \|\partial_{vt}^2 L\| T + \|\partial_{tt}^2 L\| T^2 \right) \lambda^2$$

$$+ C \|\partial_{(x,v)^2}^2 L\| \left(\|h\|^2 + |\lambda| \|h\| + \lambda^2 \right)$$

$$+ C \left(\|\partial_{xt} L\| |\lambda| T \|h\| + \|\partial_{vt} L\| (|\lambda| + \|h\|) \lambda T + \|\partial_{tt} L\| \lambda^2 T^2 \right).$$

Using $\|\partial^2 L\| \leq C$, $\|\partial_{(x,v,t)t}^2 L\| \leq \kappa$, and plug in $\Delta T = \lambda T$, we have

$$\mathbb{A}_L(\gamma_h^\lambda) - \mathbb{A}_L(\gamma)$$

$$\leq \lambda \int_0^T \left(L - \partial_v L \cdot \dot\gamma + \partial_t L \right) ds + \int_0^1 \left(\partial_x L \cdot sh + \partial_v L \cdot h \right) ds$$

$$+ C \left(\frac{1}{T} + \kappa T \right) (\Delta T)^2 + C \left(\|h\|^2 + \frac{1}{T} |\Delta T| \|h\| + \frac{1}{T^2} (\Delta T)^2 \right)$$

$$+ C\kappa \left(|\Delta T| \|h\| + (\Delta T)^2 \right)$$

Using the Euler-Lagrange equation, we have for $p(t) = \partial_v L(\gamma, \dot\gamma, t)$,

$$\frac{d}{dt}(-tH) = \frac{d}{dt}(t(L - \partial_v L \cdot \dot\gamma)) = (L - \partial_v L \cdot \dot\gamma) + t\partial_t L,$$

$$\frac{d}{dt}(p \cdot th) = \frac{d}{dt}(\partial_v L \cdot th) = \partial_v L \cdot h + \partial_x L \cdot th,$$

we get

$$\mathbb{A}_L(\gamma_h^\lambda) - \mathbb{A}_L(\gamma) \leq \lambda(-tH) \Big|_0^T + (p(T-1+t) \cdot ht) \Big|_0^1$$

$$+ C\|h^2\| + C \left(\frac{1}{T} + \epsilon \right) |\Delta T| \|h\| + C \left(\frac{1}{T} + \kappa T \right) (\Delta T)^2$$

$$\leq -H(\gamma(T), p(T), T)(\Delta T) + p(T) \cdot h + 2C\|h\|^2 + 2C \left(\frac{1}{T} + \kappa T \right) (\Delta T)^2.$$

$$\square$$

Before returning to the proof of Proposition 7.13, we state a corollary of Lemma 7.14, which gives a proof of item 4 of Proposition 6.2.

Corollary 7.15. *Let $H \in \mathbb{H}(D)$, and A be its associated action function. Then there exists a constant C depending on D such that $A(x, t_0, y, \cdot)$ is C-semi-*

concave on $(t_1 - 1, t_1 + 1)$, *and* $A(x, \cdot, y, t_1)$ *is* $C-semi\text{-}concave$ *on* $(t_0 - 1, t_0 + 1)$. *Moreover, if* $\gamma : [t_0, t_1] \to \mathbb{T}^n$ *is an extremal curve with* $t_1 - t_0 > 2$, *then*

$$-H(\gamma(t_0), p(t_0), t_0) \in -\partial_2^+ A(x, t_0, y, t_1),$$
$$-H(\gamma(t_1), p(t_1), t_1) \in \partial_4^+ A(x, t_0, y, t_1).$$

Proof. Let $T = 2$, and set $t_0' = t_1 - T$. We apply Lemma 7.14 to the curve $\gamma|[t_0', t_1]$, using the bound $\kappa \leq D$, to get

$$A(\gamma(t_0'), t_0', \gamma(t_1), t_1 + \Delta T) - A(\gamma(t_0'), t_0', \gamma(t_1), t_1) \tag{7.2}$$
$$\leq -H(\gamma(t_1), p(t_1), t_1) \cdot (\Delta T) + C(\Delta T)^2$$

for $h \in \mathbb{R}^n$ and $\Delta T \in [-1, 1]$, and some constant $C > 0$ depending only on D. Since

$$A(\gamma(t_0), t_0, \gamma(t_1) + h, t_1 + \Delta T) - A(\gamma(t_0), t_0, \gamma(t_1), t_1)$$
$$\leq A(\gamma(t_0'), t_0', \gamma(t_1) + h, t_1 + \Delta T) - A(\gamma(t_0'), t_0', \gamma(t_1), t_1),$$

we obtain (7.2) with t_0' replaced with t_0. This proves our corollary. $\square$

Proof of Proposition 7.13. Since $t_1 - t_0 > 1/\sqrt{\epsilon}$, similar to the proof of Corollary 7.15, we set $t_0' = t_1 - 1/\sqrt{\epsilon}$, and apply Lemma 7.14 to $H_\epsilon(x, p, t) = H_1(x, p) + \epsilon H_2(x, p, t)$, the curve $\gamma|[t_0', t_1]$, $T = 1/\sqrt{\epsilon}$, $\kappa = O(\epsilon)$, and $\Delta T \in [-T/2, T/2]$. We get

$$A(\gamma(t_0), t_0, \gamma(t_1) + h, t_1 + \Delta T) - A(\gamma(t_0), t_0, \gamma(t_1), t_1)$$
$$\leq A(\gamma(t_0'), t_0', \gamma(t_1) + h, t_1 + \Delta T) - A(\gamma(t_0'), t_0', \gamma(t_1), t_1)$$
$$\leq p(t_1) \cdot h - H(\gamma(t_1), p(t_1), t_1) \cdot (\Delta T) + C\|h\|^2 + C\left(\frac{1}{T} + \kappa T\right)(\Delta T)^2,$$

which implies the S_ϵ-semi-concavity of $A(x, t_0, \cdot, \cdot)$ on the set $\mathbb{R}^n \times (t_1 - T/2, t_1 + T/2)$. The semi-concavity for the first two component can be obtained by applying the same argument to the Lagrangian $L(x, -v, t)$ to the curve $\gamma(-\cdot)$ on the interval $[-t_1, -t_0]$. $\square$

Let $w(x, s)$ be a weak KAM solution to $L_\epsilon - c \cdot v$, and define $\check{w}(x, s) = \check{T}_c^{s, s+t} w(\cdot, s + t)$. Proposition 7.13 implies that w is S_ϵ-semi-concave, while $\check{w}$ is $(-S_\epsilon)$-semi-convex.

Proof of Theorem 7.10. By considering the Hamiltonian $H_\epsilon(\theta, p + c, t)$, we reduce to the case $c = 0$. This also makes the constants depend on $\|c\|$. By Proposition 6.14,

$$\phi^{-t}\mathcal{G}_{0,w} \subset \widetilde{\mathcal{I}}_{w,\check{w}}.$$

Note also for each $(x, p, s) \in \widetilde{\mathcal{I}}_{w,\check{w}}$, $-H(x, p, s) + \alpha_{L_\epsilon}(c) = \partial_s w(x, s)$. Applying Proposition 7.13 and Lemma 7.12, we obtain that for any $(x_1, p_1, t_1), (x_2, p_2, t_2) \in$

$\widetilde{\mathcal{I}}_{w,\tilde{w}}$,

$$\left\|\begin{bmatrix} p_2 - p_1 \\ H(x_2, p_2, t_2) - H(x_1, p_1, t_1) \end{bmatrix} - \begin{bmatrix} C(x_2 - x_1) \\ C\sqrt{\epsilon}(t_2 - t_1) \end{bmatrix}\right\|_{S_\epsilon^{-1}} \leq \left\|\begin{bmatrix} x_2 - x_1 \\ t_2 - t_1 \end{bmatrix}\right\|_{S_\epsilon};$$

therefore

$$\frac{1}{\sqrt{\epsilon}}|H(x_2, p_2, t_2) - H(x_1, p_1, t_1) - C\sqrt{\epsilon}(t_2 - t_1)|$$

$$\leq C\|x_2 - x_1\| + C\|t_2 - t_1\|,$$

and the proposition follows. $\qquad\square$

Chapter Eight

Cohomology of Aubry-Mather type

8.1 AUBRY-MATHER TYPE AND DIFFUSION MECHANISMS

Let $r \geq 2$,
$$H : \mathbb{T}^n \times \mathbb{R}^n \times \mathbb{T}_\varpi \to \mathbb{R}$$
be a C^r periodic Tonelli Hamiltonian in the family $\mathbb{H}(D)$.

Definition 8.1. *We say that the pair (H_*, c_*) is of Aubry-Mather type if it satisfies the following conditions:*

1. *There is an embedding*
$$\chi = \chi(x, y) : \mathbb{T} \times (-1, 1) \to \mathbb{T}^n \times \mathbb{R}^n,$$
   such that $\mathcal{C}(H) = \chi(\mathbb{T} \times (-1, 1))$ is a normally hyperbolic weakly invariant cylinder under the time-ϖ map ϕ_H. We require $\mathcal{C}(H)$ to be symplectic, i.e. the restriction of the symplectic form ω to $\overline{\mathcal{C}}$ is non-degenerate.
2. *There is $\sigma > 0$, such that the following hold for all*
$$V_\sigma(H_*) := \{\|H - H_*\|_{C^2}\} < \sigma, \quad B_\sigma(c_*) := \{\|c - c_*\| < \sigma\}.$$
   (We then say that the property below holds robustly at (H_*, c_*).)

   a) *The discrete-time Aubry set $\tilde{\mathcal{A}}^0(c) \subset \mathcal{C}(H)$. Moreover, the pull back of the Aubry set is contained in a Lipschitz graph, namely, the map*
$$\pi_x : \chi^{-1}\tilde{\mathcal{A}}_H^0(c) \subset \mathbb{T} \times (-1, 1) \to \mathbb{T}$$
   is bi-Lipschitz.
   b) *If $\chi^{-1}\tilde{\mathcal{A}}_H^0(c)$ projects onto $\mathbb{T}$, then there is a neighborhood V^0 of $\mathcal{A}_H^0(c)$ such that the strong unstable manifold $W^u(\tilde{\mathcal{A}}_H^0(c)) \cap \pi_\theta^{-1} V^0$ is a Lipschitz graph over the θ component.*

If (H_*, c_*) is of Aubry-Mather type, let $h \in \mathbb{Z}^n \simeq H_1(\mathbb{T}^n, \mathbb{Z})$ be the homology class of the curve $\chi(\mathbb{T} \times \{0\})$, called the homology class of (H_*, c_*).

Remark 8.2. The definition of Aubry-Mather type includes a much simpler case, that is when the Aubry set $\tilde{\mathcal{A}}_H(c)$ is a hyperbolic periodic orbit, still contained in a normally hyperbolic invariant cylinder (NHIC) $\mathcal{C}$. The condition 2(a) is satisfied since $\tilde{\mathcal{A}}_H^0(c)$ is discrete. Condition (b) is vacuous since the $\chi^{-1}\tilde{\mathcal{A}}_H^0(c)$ never projects onto $\mathbb{T}$. This holds, in particular, at double resonance when the energy is critical.

Definition 8.3. *We say that the pair* (H_*, c_*) *is of* bifurcation Aubry-Mather type *if there exist* $\sigma > 0$ *and open sets* $V_1, V_2 \subset \mathbb{T}^n$ *with* $\overline{V_1} \cap \overline{V_2} = \emptyset$, *and a smooth bump function*

$$f : \mathbb{T}^n \to [0, 1], \quad f|_{V_1} = 0, \quad f|_{V_2} = 1,$$

such that for all $c \in \overline{B_\sigma(c_*)}$ *and* $H \in \mathcal{V}_\sigma(H_*)$:

1. *Each of the Hamiltonians* $H^1 = H - f$ *and* $H^2 = H - (1 - f)$ *satisfies*

$$\mathcal{A}_{H^1}(c) \subset V_1 \times \mathbb{T}, \quad \mathcal{A}_{H^2}(c) \subset V_2 \times \mathbb{T}.$$

   It follows that $\tilde{\mathcal{A}}_{H^1}(c)$, $\tilde{\mathcal{A}}_{H^2}(c)$ *are both invariant sets of* H, *called the local Aubry sets.*
2. (H^1, c_*) *(resp.* (H^2, c_*)*) are of Aubry-Mather type, with the invariant cylinders* $\mathcal{C}_1$ *and* $\mathcal{C}_2$. *Moreover, they have the same homology class* h.
3. *The Aubry set*
$$\tilde{\mathcal{A}}_H(c) \subset \tilde{\mathcal{A}}_{H^1}(c) \cup \tilde{\mathcal{A}}_{H^2}(c).$$

According to item (3), the Aubry set $\tilde{\mathcal{A}}_H(c)$ may be contained in one of the local components, or both.

Remark 8.4. Suppose H is such that $\mathcal{A}_H(c)$ consists of two disjoint compact components $\mathcal{A}_H^1(c)$ and $\mathcal{A}_H^2(c)$. Choose open sets V_1, V_2 such that $\overline{V_1} \cap \overline{V_2} = \emptyset$, and $\mathcal{A}_H^1(c) \subset V_1$, $\mathcal{A}_H^2(c) \subset V_2$. Letting H^1, H^2 as in Definition 8.3, we get

$$L_{H^1} = L_H + f, \quad L_{H^2} = L_H + (1 - f).$$

Lemma 6.17 then implies $\mathcal{A}_H^j(c) = \mathcal{A}_{H^j}(c)$, $j = 1, 2$. In this sense, our definition says that each of the two local components of the Aubry set is of Aubry-Mather type.

There is another type of bifurcation in which one component of the Aubry set is of Aubry-Mather type with an invariant cylinder $\mathcal{C}_1$, and another is a hyperbolic periodic orbit. We call this the *asymmetric bifurcation*. This case appears at double resonance, when the shortest loop is *simple non-critical*. See Section 4.2.

Definition 8.5. *We say that the pair* (H_*, c_*) *is of* asymmetric bifurcation type *if there exist* $\sigma > 0$ *and open sets* $V_1, V_2 \subset \mathbb{T}^n$ *with* $\overline{V_1} \cap \overline{V_2} = \emptyset$, *and a smooth bump function*

$$f : \mathbb{T}^n \to [0, 1], \quad f|_{V_1} = 0, \quad f|_{V_2} = 1,$$

such that for all $c \in \overline{B_\sigma(c_*)}$ and $H \in V_\sigma(H_*)$:

1. *Item 1 of Definition 8.3 holds.*
2. *(H^1, c_*) is of Aubry-Mather type, with the invariant cylinder $\mathcal{C}_1$. The Aubry set $\widetilde{\mathcal{A}}_{H^2}(c_*)$ is a single hyperbolic periodic orbit.*
3. *The Aubry set*

$$\widetilde{\mathcal{A}}_H(c) \subset \widetilde{\mathcal{A}}_{H^1}(c) \cup \widetilde{\mathcal{A}}_{H^2}(c).$$

Theorem 8.6. *Suppose a pair (H_*, c_*) is of Aubry-Mather type, and let $\Gamma \ni c_*$ be a smooth curve in $\mathbb{R}^2$. Then there is $\sigma > 0$ and a residual subset $\mathcal{R}_\sigma(H_*)$ of $V_\sigma(H_*)$ such that for all $c \in \Gamma_1 := \overline{B_{\sigma_1}(c_*)} \cap \Gamma$ and $H \in \mathcal{R}_\sigma(H_*)$, at least one of the following holds:*

1. *The projected Mañé set $\mathcal{N}_H^0(c)$ is contractible as a subset of $\mathbb{T}^n$.*
2. *There is a double covering map Ξ such that the set*

$$\widetilde{\mathcal{N}}_{H \circ \Xi}^0(\xi^* c) \setminus \Xi^{-1} \widetilde{\mathcal{N}}_H^0(c)$$

is totally disconnected.

Theorem 8.7. *Suppose (H_*, c_*) is of either*

- *bifurcation AM type, or*
- *asymmetric bifurcation type,*

and $c_* \in \Gamma$, where Γ is a smooth curve in $\mathbb{R}^2$, then there is $\sigma > 0$ and residual $\mathcal{R} \subset V_\sigma(H_*)$ such that for $c \in B_\sigma(c_*) \cap \Gamma$ and $H \in \mathcal{R}$, at least one of the following holds:

1. *$\widetilde{\mathcal{A}}_H^0(c)$ has a unique static class, (hence $\widetilde{\mathcal{A}}_H^0(c) = \widetilde{\mathcal{N}}_H^0(c)$), and either*

    a) *the projected Mañé set $\mathcal{N}_H^0(c)$ is contractible as a subset of $\mathbb{T}^n$, or*
    b) *there is a double covering map Ξ such that the set*

$$\widetilde{\mathcal{N}}_{H \circ \Xi}^0(\xi^* c) \setminus \Xi^{-1} \widetilde{\mathcal{N}}_H^0(c)$$

is totally disconnected.
2. *$\widetilde{\mathcal{A}}_H^0(c)$ has two static classes, and*

$$\widetilde{\mathcal{N}}_H^0(c) \setminus \widetilde{\mathcal{A}}_H^0(c)$$

is a discrete set.

Analogous result in the a priori unstable setting is due to Mather [59] and Cheng-Yan [26, 27]. Our definition is more general and applies, as will be seen, to both single- and double-resonant settings. The prove is based on the result of [13], which applies to our setting with appropriate changes. Here we describe the changes needed for the proof in [13] to apply.

If (H_*, c_*) is of Aubry-Mather type, let $h \in \mathbb{Z}^n \simeq H_1(\mathbb{T}^n, \mathbb{Z})$ be its homology class. Then any minimal measure contained in $\tilde{\mathcal{A}}^0(c)$ must have rotation vector λh, $\lambda \in \mathbb{R}$. Moreover, similar to the case of twist map, all such measures have the same rotation vector λh. We say the rotation vector is rational/irrational if λ is rational/irrational.

The following proposition is a consequence of the Hamiltonian Kupka-Smale theorem, see for example [71].

Proposition 8.8. *There are $\sigma_1, \sigma_2 > 0$, and a residual subset $\mathcal{R}_1$ of the ball $V_{\sigma_2}(H_*)$, such that for all $c \in \Gamma_1 := \overline{B_{\sigma_1}(c^*)} \cap \Gamma$, $H \in \mathcal{R}_1$, if c has rational rotation vector, then $\pi \chi^{-1} \tilde{\mathcal{A}}^0(c) \neq \mathbb{T}$.*

Let $1 \leq j \leq n$, and e_j be the jth coordinate vector in $\mathbb{R}^n$. Suppose that $e_j \nmid h$; define the covering map $\xi : \mathbb{T}^n \to \mathbb{T}^n$ by

$$\xi(\theta_1, \cdots, \theta_n) = (\theta_1, \cdots, 2\theta_j, \cdots, \theta_n).$$

- For $H \in \mathcal{R}_1$, define

$$\Gamma_*(H) = \{c \in \Gamma_1 : \pi\chi^{-1}(\mathcal{N}_H^0(c)) = \mathbb{T}\}. \qquad (8.1)$$

Proposition 8.8 implies that each $c \in \Gamma_*(H)$ has an irrational rotation vector, and the Aubry set has a unique static class, and $\mathcal{A}_H^0(c) = \mathcal{N}_H^0(c)$. Each $\Gamma_*(H)$ is compact due to upper semi-continuity of the Mañé set.
- For $H \in \mathcal{R}_1$ and $c \in \Gamma_*(H)$, the lifted Aubry set $\tilde{\mathcal{A}}_{H \circ \Xi}^0(\xi^* c) = \Xi^{-1} \tilde{\mathcal{A}}_H(c)$ has two static classes, denote them $\tilde{\mathcal{S}}_1, \tilde{\mathcal{S}}_2$ (and projections $\mathcal{S}_1, \mathcal{S}_2$). We have the decomposition

$$\tilde{\mathcal{N}}_{H \circ \Xi}^0(\xi^* c) = \tilde{\mathcal{S}}_1 \cup \tilde{\mathcal{S}}_2 \cup \tilde{\mathcal{H}}_{12} \cup \tilde{\mathcal{H}}_{21},$$

where $\tilde{\mathcal{H}}_{ij}$ consists of heteroclinic orbits from $\tilde{\mathcal{S}}_i$ to $\tilde{\mathcal{S}}_j$. We will use the notation $\tilde{\mathcal{S}}_i(H, c), \tilde{\mathcal{H}}(H, c)$ to show dependence on the pair (H, c), and $\mathcal{H}_{ij}$ for projection of $\tilde{\mathcal{H}}_{ij}$.
- For $H \in \mathcal{R}_1$ and $c \in \Gamma_*(H)$, consider the (discrete-time) Peierl's barrier (see (6.6))

$$h(\zeta_1, \cdot), \ h(\zeta_2, \cdot), \quad h(\cdot, \zeta_1), \ h(\cdot, \zeta_2), \quad \zeta_i \in \mathcal{S}_i, \ i = 1, 2,$$

where $h = h_{H \circ \Xi, \xi^* c}$. These functions are independent of the choice of ζ_i except for an additive constant. We define

$$b_{H,c}^-(\theta) = h(\zeta_1, \theta) + h(\theta, \zeta_2) - h(\zeta_1, \zeta_2)$$

and $b_{H,c}^+$ by switching ζ_1, ζ_2. The functions $b_{H,c}^\pm$ are non-negative and vanish on $\mathcal{H}_{12} \cup \mathcal{S}_1 \cup \mathcal{S}_2$ and $\mathcal{H}_{21} \cup \mathcal{S}_1 \cup \mathcal{S}_2$, respectively.
- Consider small neighborhoods V_1, V_2 of $\mathcal{S}_1(H_*, c_*), \mathcal{S}_2(H_*, c_*)$, and define $K = \mathbb{T}^n \setminus (V_1 \cup V_2)$. By semi-continuity of the Aubry set, for sufficiently small

σ_1, σ_2, K is disjoint from $\mathcal{S}_i(H, c)$ for all $c \in \Gamma_1$ and $H \in \mathcal{R}_1$. Moreover, $\pi^{-1}K$ intersects every orbit of $\widetilde{\mathcal{H}}_{12}(H, c)$ and $\widetilde{\mathcal{H}}_{12}(H, c)$.

Lemma 8.9 (Lemma 5.2 of [13]). *For each $(H, c) \in \mathcal{R}_1 \times \Gamma_1$, the set*

$$\widetilde{\mathcal{N}}_{H \circ \Xi}(\xi^* c) \setminus \Xi^{-1} \widetilde{\mathcal{N}}_H(c)$$

is totally disconnected if and only if

$$\mathcal{N}_{H \circ \Xi}(\xi^* c) \cap K = (\mathcal{H}_{12} \cup \mathcal{H}_{21}) \cap K$$

is totally disconnected.

We will show that the set of $H \in \mathcal{R}_1$ with the following property contains a dense G_δ set: for each $c \in \Gamma_*(H)$, $\mathcal{N}_{H \circ \Xi}(\xi^* c) \cap K$ is totally disconnected. The following lemma implies the G_δ property.

Lemma 8.10 (Lemma 5.3 of [13]). *Let $K \subset \mathbb{T}^n$ be compact, then the set of $H \in \mathcal{R}_1$ such that for all $c \in \Gamma_*(H)$, the set $\mathcal{N}_{H \circ \Xi}(\xi^* c)$ is totally disconnected, is a G_δ set.*

The following proposition allows local perturbations of $b^\pm_{H,c}$ simultaneously for all c's in a small ball. Let $B_\sigma(x)$ denote the ball of radius σ at x in a metric space.

Proposition 8.11 (Proposition 5.2 of [13]). *Let $H_* \in \mathcal{R}_1$, $c_* \in \Gamma_*(H)$, and $K \cap \mathcal{A}_{H \circ \Xi}(\xi^* c) = \emptyset$. Then there is $\sigma > 0$ such that for all*

$$H \in \mathcal{R}_1 \cap B_\sigma(H_*), \quad \theta_0 \in K \cap \mathcal{H}_{12}(H_*, c_*), \quad \varphi \in C_c^r(B_\sigma(\theta_0)) \text{ with } \|\varphi\|_{C^r} < \sigma,$$

there is a Hamiltonian H_φ such that:

1. *For all $c \in B_\sigma(c_*)$, the Aubry sets $\widetilde{\mathcal{A}}_{H_\varphi \circ \Xi}(\xi^* c)$ coincides with $\widetilde{\mathcal{A}}_{H \circ \Xi}(\xi^* c)$ with the same static classes. In particular, $B_\sigma(c_*) \cap \Gamma_*(H) = B_\sigma(c_*) \cap \Gamma_*(H_\varphi)$.*
2. *For all $c \in B_\sigma(c_0) \cap \Gamma_*(H)$, there exists a constant $e \in \mathbb{R}$ such that*

$$b^+_{H_\varphi, c}(\theta) = b^+_{H, c}(\theta) + \varphi(\theta) + e, \quad \theta \in B_\sigma(\theta_0). \tag{8.2}$$

The same holds for $\theta_0 \in K \cap \mathcal{H}_{21}(H_*, c_*)$, with b^+ replaced with b^- in (8.2). Moreover, for each $H \in \mathcal{R}_1 \cap B_\sigma(H_*)$, $\|H_\varphi - H\|_{C^r} \to 0$ when $\|\varphi\|_{C^r} \to 0$.

We will use Proposition 8.11 to locally perturb the functions $b^+_{H,c}$, therefore, perturbing

$$\mathcal{H}_{12}(H, c) \cap B_\sigma(\theta_0) = \operatorname{argmin} b^+_{H,c} \cap B_\sigma(\theta_0).$$

Similarly for b^-. However, as observed by Mather and Cheng-Yan, this requires additional information on how $b^\pm_{H,c}$ depends on c.

Proposition 8.12 (Section 8.3). *There is $\beta > 0$ such that for each $H \in \mathcal{R}_1$, the maps $c \mapsto b_{H,c}^{\pm}$ from $\Gamma_*(H)$ to $C^0(\mathbb{T}^n, \mathbb{R})$ are β-Hölder.*

As a result, the set $\{b_{H,c}^{\pm} : c \in \Gamma^*(H)\}$ has Hausdorff dimension at most $1/\beta$ in $C^0(\mathbb{T}^n, \mathbb{R})$. The following lemma allows us to take advantage of this fact.

Lemma 8.13 (Lemma 5.6 of [13]). *Let $\mathcal{F} \subset C^0([-1, 1]^n, \mathbb{R})$ be a compact set of finite Hausdorff dimension. The following property is satisfied on a residual set of functions $\varphi \in C^r(\mathbb{R}^n, \mathbb{R})$ (with the uniform C^r norm):*

For each $f \in \mathcal{F}$, the set of minima of the function $f + \varphi$ on $[-1, 1]^n$ is totally disconnected.

As a consequence, for each open neighborhood Ω of $[-1, 1]^n$ in $\mathbb{R}^n$, there exist arbitrarily C^r-small compactly supported functions $\varphi : \Omega \to \mathbb{R}$ satisfying this property.

Proof of Theorem 8.6. Let (H_*, c_*) be of Aubry-Mather type, and let $\sigma > 0$ be as in Definition 8.1. Let K be as in Lemma 8.9 and denote $\Gamma_1 = \overline{B_\sigma(c_*)} \cap \Gamma$. Let $\mathcal{R}_2 \subset \mathcal{R}_1 \cap \mathcal{V}_\sigma(H_*)$ be the set of Hamiltonians such that for all $c \in \Gamma_1 \cap \Gamma_*(H)$, the set $\tilde{\mathcal{N}}_{H \circ \Xi}^0(\xi^* c) \setminus \tilde{\mathcal{N}}_H^0(c)$ is totally disconnected. According to Lemma 8.10, this set is G_δ, we will to show that it is dense on $\mathcal{V}_{\sigma_2}(H_*)$ for some $0 < \sigma_2 < \sigma$.

Consider $c \in \Gamma_1 \cap \Gamma_*(H_*)$, let $\sigma_c > 0$ be small so that Proposition 8.11 applies to the pair (H_*, c) on the set K. For each $\theta_0 \in \tilde{\mathcal{N}}_{H_* \circ \Xi}^0(\xi^* c) \cap K$, define

$$D_{\sigma_c}(\theta_0) = \{\theta : \max_i |\theta^i - \theta_0^i| \leq \sigma_c/(2\sqrt{n})\} \subset B_{\sigma_c}(\theta_0).$$

Proposition 8.12 implies that the family of functions

$$b_{H,c}^{\pm}, \quad c \in \Gamma_1 \cap \Gamma_*(H)$$

has Hausdorff dimension at most $1/\beta$, therefore we can apply Lemma 8.13 on the cube $D_{\sigma_c}(\theta_0)$ for each $H \in \mathcal{R}_1$. We find arbitrarily small functions φ compactly supported in $D_{\sigma_c}(\theta_0)$ and such that each of the functions

$$b_{N,c}^{\pm} + \varphi, \quad c \in \Gamma_1 \cap \Gamma^*(H)$$

have a totally disconnected set of minima in $D_{\sigma_c}(\theta_0)$. We then apply Proposition 8.11 to get Hamiltonians H_φ approximating H. We obtain:

- The set of Hamiltonians H such that $\mathcal{N}_{H \circ \Xi}(\xi^* c) \cap D_{\sigma_c}(\theta_0)$ is totally disconnected for each $c \in \Gamma_*(H)$ is dense in $\mathcal{R}_1 \cap \mathcal{V}_{\sigma_c}(H_*)$. By Lemma 8.10, it is also G_δ, and therefore residual.

Since K is compact, there is a finite cover $K \subset \bigcup_{i=1}^k D_{\sigma_i}(\theta_i)$, such that

- For a residual set $\mathcal{R}^i(c)$ of $H \in B_{\sigma_c}(H_*)$, the set $\mathcal{N}_{H \circ \Xi}(\xi^* c) \cap D_{\sigma_c}(\theta_i)$ is totally disconnected for all $i = 1, \ldots, k$ and $c \in \Gamma_*(H)$.

Take $\mathcal{R}_c = \bigcap_i \mathcal{R}^i$, then for $H \in \mathcal{R}_c$, the set $\mathcal{N}_{H \circ \Xi}(\xi^* c) \cap K$ is totally disconnected. Finally, we consider a finite covering $\Gamma_1 \subset \bigcup_j B_{\sigma_{c_j}}(c_j)$ and repeat the preceding argument. $\qquad\qquad\qquad\qquad\qquad\qquad\qquad\qquad\qquad\qquad\square$

The next two sections are dedicated to proving Proposition 8.12.

8.2   WEAK KAM SOLUTIONS ARE UNSTABLE MANIFOLDS

An important consequence of the Aubry-Mather type is that the local unstable manifold coincides with an elementary weak KAM solution.

Proposition 8.14. *Suppose (H_*, c_*) is of Aubry-Mather type. Then for each $(H, c) \in \mathcal{V}_\sigma(H_*) \times B_\sigma(c_*)$ such that $\chi^{-1} \widetilde{\mathcal{A}}_H^0(c)$ projects onto $\mathbb{T}$, we have*

$$W^u(\widetilde{\mathcal{A}}_H^0(c)) \cap \pi_\theta^{-1} V^0 = \{(\theta, c + \nabla u(\theta)) : \quad \theta \in V^0 \cap \mathcal{A}_H^0(c)\},$$

where $u(\theta) = h_{H,c}(\zeta, \theta)$ for some $\zeta \in \mathcal{A}_H^0(c)$.

Lemma 8.15. *Suppose $\widetilde{\mathcal{A}} \subset \mathbb{T}^n \times \mathbb{R}^n$ is a closed 1-dimensional Lipschitz submanifold, invariant under ϕ_H, partially hyperbolic, and that $W^u(\widetilde{\mathcal{A}})$ is a Lipschitz manifold in $\mathbb{R}^n \times \mathbb{T}^n$. Then $W^u(\widetilde{\mathcal{A}})$ is isotropic in the sense that for Lebesgue almost every $z \in W^u(\widetilde{\mathcal{A}})$ and every $v, w \in T_z W^u(\widetilde{\mathcal{A}})$, we have*

$$(d\theta \wedge dp)(v, w) = 0. \tag{8.3}$$

If $W^u(\widetilde{\mathcal{A}})$ is a Lipschitz graph over a neighborhood $V^0 \subset \mathbb{R}^n$, then there exists a $C^{1,1}$ function $u : V^0 \to \mathbb{R}$ such that

$$W^u(\widetilde{\mathcal{A}}) = \{(\theta, \nabla u(\theta)) : \theta \in V^0\}.$$

Proof. Denote $W^u = W^u(\widetilde{\mathcal{A}})$, $\phi = \phi_H$ and $\omega = d\theta \wedge dp$. Since W^u is backward invariant, and every ϕ_H backward orbit is asymptotic to $\widetilde{\mathcal{A}}$, it suffices to prove (8.3) for $v, w \in T_z W^u$, for a.e. $z \in W^u$ in a neighborhood U of $\widetilde{\mathcal{A}}$. Since the partially hyperbolic splitting of $\widetilde{\mathcal{A}}$ extends to a neighborhood, for $z \in U \cap W^u$, there is a continuous splitting

$$T_z W^u = E^c(z) \oplus E^u(z),$$

such that for any $v \in T_z W^u$ and $v^u \in E^u(z)$, we have $\|(D\phi_H^{-k})v\| \leq C\mu^k$ and $\|(D\phi_H^{-k})v^u\| \leq C\lambda^k$ for some $C > 0$ and $0 < \mu < \lambda^{-1}$. Moreover, $\dim E^c = 1$ since $\widetilde{\mathcal{A}}$ is 1-dimensional. For a.e. $z \in U \cap W^u$ such that $T_z W^u$ is well defined:

- If $v, w \in E^c(z)$, then $\omega(v, w) = 0$ since $E^c(z)$ is 1-dimensional.

- If $v \in T_z W^u$ and $w \in E^u(z)$, then

$$\omega_z(v, w) = \omega_{\phi^{-k}}((D\phi^{-k}(z))v, (D\phi^{-k}(z))w) \le C^2(\lambda\mu)^k \to 0, \quad \text{as } k \to \infty.$$

This implies $\omega(v, w) = 0$ for all $v, w \in T_z W^u$.

Suppose now that W^u is a graph over V^0, then W^u is Lagrangian. Fix $z_0 \in \tilde{\mathcal{A}}$, for any $z \in W^u$, consider a curve $\zeta(t) = (\theta, p)(t)$ such that $\zeta(0) = z_0$, $\zeta(t) = z$, then the path integral $\int_0^t p(t) \cdot \dot{\theta}(t) dt$ is independent of the path and defines a function $u : V^0 \to \mathbb{R}$. For a different choice of z_0, u differs by a constant. By differentiating the integral, we see that $(\theta, p) \in W^u \cap V^0$ if and only if $p = \nabla u(\theta)$.   □

Lemma 8.16. *Under the same assumption as Proposition 8.14, consider the time-periodic Aubry set* $\tilde{\mathcal{A}}_H(c) = \bigcup_{t \in [0, \varpi(H)]} \phi_H^t(\tilde{\mathcal{A}}_H^0(c)) \subset \mathbb{T}^d \times \mathbb{R}^d \times (\mathbb{R}/\varpi\mathbb{Z})$ *and* $W^u(\tilde{\mathcal{A}}_H(c))$ *its strong unstable manifold. Then there exists a neighborhood* V *of* $\mathcal{A}_H(c)$ *and a* $C^{1,1}$ *function* $w : V^0 \times \mathbb{R} \to \mathbb{R}$, $w(\cdot, t + \varphi(H)) = w(\cdot, t)$, *such that* w *solves the Hamilton-Jacobi equation*

$$w_t + H(\theta, c + \partial_\theta w(\theta, t), t) = \alpha_H(c),$$

and $(\theta, \partial_\theta w(\theta, t), t) = W^u(\tilde{\mathcal{A}}_H(c))$ *for all* $(\theta, t) \in V$.

Assume, in addition, that $\tilde{\mathcal{A}}_H(c)$ *has a unique static class. Then there is* $C \in \mathbb{R}$ *and* $(\zeta, \tau) \in \mathcal{A}_H(c)$ *such that*

$$w(\theta, t) = h_{H,c}(\zeta, \tau, \theta, t) + C, \quad (\theta, t) \in V.$$

Proof. By considering $H(\theta, c + p, t)$, we reduce to the case $c = 0$. Denote $\tilde{\mathcal{A}}^0 = \tilde{\mathcal{A}}_H^0(0)$, $\tilde{\mathcal{A}}^t = \phi_H^t(\tilde{\mathcal{A}}^0)$, $W^{u,0} = W^u(\tilde{\mathcal{A}}^0)$, $W^{u,t} = \phi^t(W^u)$, $W^u = W^u(\tilde{\mathcal{A}}) \subset \mathbb{T}^d \times \mathbb{R}^d \times (\mathbb{R}/\varpi\mathbb{Z})$. Let $w_0 : \mathbb{T}^d \times \mathbb{R} \to \mathbb{R}$ denote a weak KAM solution of H. We have:

- For each $(\theta, p, t) \in \tilde{\mathcal{A}}^t$, $p = \partial_\theta w_0(\theta, t)$.
- Applying Lemma 8.15 to each $\tilde{\mathcal{A}}^t$ with $t \in [0, \varpi(H)]$, we get that each $W^{u,t}$ is the gradient of a function $u^t(\theta)$. Moreover, u^t is uniquely determined by requiring $u^t(\theta) = w_0(\theta, t)$ for $\theta \in \pi_\theta \tilde{\mathcal{A}}^t$. Write $w(\theta, t) = u^t(\theta)$, then $(\theta, \partial_\theta w(\theta, t), t)$ coincides with $W^u(\tilde{\mathcal{A}})$ in a neighborhood

Let V be a neighborhood of $\tilde{\mathcal{A}}_H(0)$ contained in the projection $\pi_\theta W^u(\tilde{\mathcal{A}}) \subset \mathbb{T}^n \times (\mathbb{R}/\varphi\mathbb{Z})$. Then for each $(\theta, t) \in V$, we have $(\theta(s), p(s), s) = \phi_H^s(\theta, \partial_\theta w(\theta, t), t) \in W^u$ for all sufficiently small s. It follows that

$$\dot{\theta} = \partial_p H(\theta, \partial_\theta w(\theta, t), t), \quad \frac{d}{dt}(\partial_\theta w(\theta(t), t)) = -\partial_\theta H(\theta, \partial_\theta w(\theta, t), t).$$

Suppose $(\theta(t), t)$ is a point at which $\partial_{\theta\theta} w(\theta, t)$ exists (such points have full

Lebesgue measure), then

$$\partial_{\theta,t} w(\theta,t) + \partial_{\theta\theta} w(\theta,t)\dot{\theta}(t) = -\partial_{\theta} H(\theta, \partial_{\theta} w(\theta,t), t).$$

Combine the last two equations, we get

$$0 = \partial_{\theta t} w + \partial_{\theta} H + \partial_{\theta\theta} w \partial_p H = \partial_{\theta}\left(\partial_t w(\theta,t) + H(\theta, \partial_{\theta} w(\theta,t), t), t)\right).$$

Since this equality holds for a.e (θ,t), we conclude that there is a function $g(t) : (\mathbb{R}/\varpi\mathbb{Z}) \to \mathbb{R}$ such that

$$w_t + H(\theta, c + \partial_{\theta} w(\theta,t), t) = g(t).$$

Finally, note that $g(t) = \alpha_H(0)$ on $\tilde{\mathcal{A}}$, and we conclude that $g(t) = \alpha_H(0)$.

We now prove w coincides with a Peierl's barrier function. According to Proposition 4.1.8 of [37], when u solves the Hamilton-Jacobi equation, u is dominated by L, and that for each $(\theta,t) \in V$ there exists $\epsilon > 0$ and a curve $\gamma : [t - \epsilon, t] \to \mathbb{T}^d$ such that γ is calibrated by w (see Section 6.2). Theorem 4.2.6 of [37] implies that the orbit $(\gamma(s), \partial_{\theta} w(\gamma(s), s), s)$ solves the Hamiltonian equation, and in particular, $(\gamma(s), s) \in V$ for all $s < t$. Repeating the argument, we conclude that at each (θ,t) there exists a backward calibrated curve $\gamma : (-\infty, t] \to \mathbb{T}^d$ such that $\gamma(t) = \theta$. We note that the associated Hamiltonian orbit is in the unstable manifold of $\tilde{\mathcal{A}}$.

Suppose $T_n < 0$ is such that $(\gamma(-T_n), -T_n) \to (\zeta, \tau) \in \mathcal{A}_H(0)$. Note that $z' = (\zeta, \partial_{\theta} w(\zeta, \tau)\tau)$ is the unique point such that $(\theta, \partial_{\theta} w(\theta,t), t) \in W^u(z')$. Let $\gamma : (-\infty, t] \to \mathbb{T}^d$ the backward calibrated orbit ending at (θ,t). Using the uniform Lipschitz property of the action function A_H, we have

$$w(\theta,t) = \lim_{n\to\infty} w(\gamma(t - T_n), t - T_n) + A_H(\gamma(t - T_n), t - T_n, \theta, t)$$

$$= w(\zeta, \tau) + \lim_{n\to\infty} A_H(\zeta, t - T_n, \theta, t) \geq w(\zeta, \tau) + h_H(\zeta, \tau, \theta, t).$$

On the other hand, for arbitrary $T_n' = \varpi n + t - \tau$, by the domination property

$$w(\theta,t) \leq w(\zeta, \tau) + A_H(\zeta, \tau - T_n, \theta, t),$$

we get $w(\theta,t) \leq w(\zeta, \tau) + h_H(\zeta, \tau, \theta, t)$ by taking lim inf to both sides. Note that we have not proven what's needed since (ζ, τ) depends on (θ,t). When there is only one static class we have

$$h_H(\zeta_1, \tau_1, \theta, t) = h_H(\zeta_1, \tau_1, \zeta_2, \tau_2) + h_H(\zeta_2, \tau_2, \theta, t), \quad (\zeta_1, \tau_1), (\zeta_2, \tau_2) \in \mathcal{A}_H(0),$$

which allows a consistent choice of (ζ, τ) for all (θ,t).   □

Proof of Proposition 8.14. By converting Definition 8.1, (2)(b), to its continuous counterpart, the condition of Lemma 8.16 is satisfied for $\tilde{\mathcal{A}}_H(c)$. The proposition follows by taking the zero section of the weak KAM solution.   □

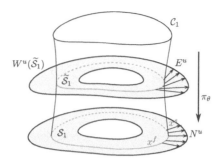

Figure 8.1: Local coordinates in the configuration space near $\mathcal{S}_1$.

8.3   REGULARITY OF THE BARRIER FUNCTIONS

In this section we prove Proposition 8.12. Let $c_* \in \Gamma_*(H_*)$, $H \in \mathcal{V}_\sigma(H_*)$, and $c \in B_\sigma(c_*) \cap \Gamma_*(H)$, then according to our assumption, the Aubry set $\widetilde{\mathcal{A}}_H(c)$ is contained in the cylinder $\mathcal{C}$ and has a unique static class.

Let us now consider the lift $\widetilde{\mathcal{A}}^0_{H \circ \Xi}(\xi^* c)$, which has two components $\widetilde{\mathcal{S}}_1, \widetilde{\mathcal{S}}_2$. The cylinder $\mathcal{C}$ lifts to two disjoint cylinders $\mathcal{C}_1, \mathcal{C}_2$. For the rest of the discussion, we will consider only the static class $\widetilde{\mathcal{S}}_1$ as the other case is similar.

Definition 8.1 ensures that there is a Lipschitz function $y = g(x) \in (-1, 1)$, $x \in \mathbb{T}$, such that

$$\widetilde{\mathcal{S}}_1(H, c) = \{\chi(x, g(x)) : x \in \mathbb{T}\} =: \{F_c(x) = (F_c^\theta(x), F_c^p(x)) : x \in \mathbb{T}\}$$
$$\subset \mathbb{T}^n \times \mathbb{R}^n.$$

Now suppose $c, c' \in \Gamma_*(H) \cap B_\sigma(c_*)$. First we have:

Lemma 8.17. *There is $C > 0$ such that*

$$\sup_x \|F_c(x) - F_{c'}(x)\| \le C\|c - c'\|^{\frac{1}{2}}.$$

Proof. The proof is the same as the one in Lemma 5.8, [13] where the proof only used the symplecticity of the cylinder. The main idea is that $\sup_x \|F_c(x) - F_{c'}(x)\|$ is $\frac{1}{2}$ Hölder with respect to the area between the two invariant curves restricted to the cylinder, while the latter is equivalent to the symplectic area by assumption. A direct calculation shows the symplectic area is bounded by $\|c' - c\|$. $\qquad\square$

Let us now denote

$$u_c(\theta) = h_{H \circ \Xi, \xi^* c}(\zeta_1, \theta), \quad \zeta_1 \in \mathcal{S}_1(H, c).$$

Before moving forward, we define a convenient local coordinate system near $\mathcal{S}_1$. According to Proposition 8.14, the graph $(\theta, c + \nabla u_c(\theta))$ locally coincides with $W^u(\widetilde{\mathcal{S}}_1(H,c))$. The (strong) unstable bundle $E^u(z)$ of $z \in \widetilde{\mathcal{S}}_1(H,c)$ is tangent to $W^u(\widetilde{\mathcal{S}}_1(H,c))$ at every point and transverse to the tangent cone of $\widetilde{\mathcal{S}}_1(H,c)$. Since $W^u(\widetilde{\mathcal{S}}_1(H,c))$ is a Lipschitz graph over $\mathbb{T}^n$, the projection $N^u(z) := d\pi_\theta E^u(z)$ forms a non-zero section of the normal bundle to $\mathcal{S}_1(H,c) = \pi_\theta \widetilde{\mathcal{S}}_1(H,c)$ within the configuration space $\mathbb{T}^n$. By choosing an orthonormal basis $e_1(z), \cdots, e_{n-1}(z)$, $N^u(z)$ naturally defines a coordinate system $(x^f, x^s) \in \mathbb{T} \times \mathbb{R}^{n-1}$ on the tubular neighborhood of $\mathcal{S}_1$:

$$\iota(x^f, x^s) = F_c^\theta(x^f) + \sum_{i=1}^{n-1} e_i(F_c^\theta(x^f))x_i^s.$$

See Figure 8.1. We note that the coordinate system is only Hölder since the unstable bundle is only Hölder a priori. Therefore, we consider C^∞ functions that approximate F_c^θ and e_i in the C^0 sense, the new coordinate system is still well defined near $\mathcal{S}_1$, and the x^s coordinate projects onto the unstable direction. In the sequel, we fix such a coordinate system using $W^u(\widetilde{\mathcal{S}}_1(H_*, c_*))$. Due to semi-continuity, for (H,c) close to (H_*, c_*), the coordinate system is defined in a neighborhood of $\mathcal{S}_1(H,c)$.

For a partially hyperbolic invariant set, the holonomy map for the strong unstable foliation is a priori only Hölder (see [70]). The Hölder constant depends on the ratio between the unstable expansion rate and norm of the tangent map in the center direction. Assume that $\beta_0 \in (0,1]$ is the Hölder exponent for the strong unstable foliation of $W^u(\widetilde{\mathcal{S}}_1(H,c))$. We obtain the following regularity result:

Lemma 8.18. *There is $\sigma \in (0,1)$ such that for $H \in \mathcal{V}_\sigma(H_*)$, $c, c' \in \Gamma_*(H) \cap B_\sigma(c_*)$, $\theta \in B_\sigma(\mathcal{S}_1(H,c_*))$, there is $\beta > 0$, $C_1 > 0$, $C_2 \in \mathbb{R}$ such that*

1. *$|\nabla u_c(\theta) - \nabla u_{c'}(\theta)| \leq C_1 \|c - c'\|^\beta$;*
2. *$|u_c(\theta) - u_{c'}(\theta) - C_2| \leq C_1 \|c - c'\|^\beta$.*

Moreover, the same holds with $\mathcal{S}_1$ replaced with $\mathcal{S}_2$.

Proof. The proof follows essentially Lemma 5.9 in [13]. Let σ_1 be small enough such that the local coordinates (x^f, x^s) are defined for $|x^s| < \sigma$. We then consider the weak KAM solution $u_c \circ \iota(x^f, x^s)$ instead, which we still denote as $u_c(x^f, x^s)$, abusing the notation.

Let $|x^s| < \sigma_1$, let $y = (x^f, x^s, c + \nabla u_c(x^f, x^s))$, and let $z \in \mathcal{S}_1(H,c)$ be such that $y \in W^u(z)$. We then define $z' \in \mathcal{S}_1(H,c)$ to be the unique such point with $x^f(z') = x^f(z)$. Finally define $y' \in W^u(z')$ to be such that $x^s(y') = x^s(y)$, which is possible since $W^u(z')$ is a graph over the x^s coordinates. See Figure 8.2.

We note that within the center unstable manifold $W^u(\mathcal{C}_1)$, the NHIC $\mathcal{C}_1$ on one hand, and $x^s = x^s(y)$ on the other hand serve as two transversals to the strong unstable foliation $\{W^u(\cdot)\}$. Since the foliation is β_0-Hölder, there exists

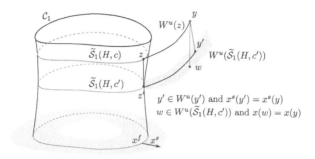

$$C_1$$
$$\widetilde{\mathcal{S}}_1(H,c)$$
$$\widetilde{\mathcal{S}}_1(H,c')$$
$$W^u(z)$$
$$y$$
$$y'$$
$$z$$
$$w$$
$$W^u(\widetilde{\mathcal{S}}_1(H,c'))$$
$$z'$$
$$y' \in W^u(y') \text{ and } x^s(y') = x^s(y)$$
$$w \in W^u(\widetilde{\mathcal{S}}_1(H,c')) \text{ and } x(w) = x(y)$$
$$x^f \quad x^s$$

Figure 8.2: Proof of Lemma 8.18.

$C > 0$ (throughout the proof, C denotes a generic constant) such that

$$\|y - y'\| \le C\|z - z'\|^{\beta_0} \le C\|c - c'\|^{\frac{\beta_0}{2}}.$$

Denoting $w = (\theta, c' + \nabla u_{c'}(\theta))$ and noting $y' \in W^u(\mathcal{S}_1(H,c')) = \{(\theta, \nabla u_{c'}(\theta))\}$ which is locally a C^1 graph, we get for $C > 0$

$$\|w - y'\| \le C\|\pi_\theta(w) - \pi_\theta(y')\| = C\|\pi_\theta(y) - \pi_\theta(y')\| \le C\|y - y'\|,$$

therefore

$$\|c + \nabla u_c(\theta) - c' - \nabla u_{c'}(\theta)\| \le \|w - y\| \le \|w - y'\| + \|y - y'\|$$
$$\le C\|y - y'\| \le C\|c - c'\|^{\frac{\beta_0}{2}}.$$

Since $\|c - c'\| < \sigma < 1$, $\|c - c'\| < \|c - c'\|^{\beta_0/2}$, we have

$$\|\nabla u_c(\theta) - \nabla u_{c'}(\theta)\| \le \|c - c'\| + C\|c - c'\|^{\frac{\beta_0}{2}} \le (C+1)\|c - c'\|^{\frac{\beta_0}{2}}.$$

We now revert the local coordinate ι to obtain item 1 with $\beta = \beta_0/2$, and possibly changing σ and C_1. Item 2 is obtained from item 1 by direct integration. $\square$

8.4 BIFURCATION TYPE

We prove Theorem 8.7 in this section. Suppose (H_*, c_*) is of bifurcation type, and let H^1, H^2 be the Hamiltonians as defined in Definition 8.3, then there exist cylinders $\mathcal{C}_1, \mathcal{C}_2$ containing the local Aubry sets $\widetilde{\mathcal{A}}_{H^1}(c)$ and $\widetilde{\mathcal{A}}_{H^2}(c)$.

Let us denote $\alpha_H^i(c) = \alpha_{H^i}(c)$, called the local alpha functions. Then $\widetilde{\mathcal{A}}_H(c)$ has two static classes if and only if $\alpha_H^1(c) = \alpha_H^2(c)$. Moreover, for each c, the rotation vector $\rho_{H^i}(c)$ is uniquely defined, as a result, α_H^i is a C^1 function.

Proposition 8.19. *Let H_*, c_* be of bifurcation Aubry-Mather type. There is $\sigma > 0$ such that for an open and dense subset of $H \in \mathcal{V}_\sigma(H_*)$, there are at most finitely many $c \in \overline{B_\sigma}(c_*) \cap \Gamma$ for which $\alpha_H^1(c) = \alpha_H^2(c)$.*

For each such (bifurcation) c, we have $\tilde{\mathcal{A}}_{H^1}^0(c), \tilde{\mathcal{A}}_{H^2}^0(c)$ both supported on hyperbolic periodic orbits. Moreover, $\tilde{\mathcal{N}}_H^0(c) \setminus \tilde{\mathcal{A}}_H^0(c)$ is a discrete set.

Proof. Let σ be as in Definition 8.3. For the rest of the proof, we refer to $\overline{B_\sigma}(c_*) \cap \Gamma$ as Γ to simplify notations.

Consider the family of Hamiltonians $H_\lambda = H(\theta, p, t) - \lambda(1 - f)$ (f is the mollifier function in Definition 8.3). Lemma 6.17 implies

$$\alpha_{H_\lambda}^1(c) = \alpha_H^1(c) + \lambda, \quad \alpha_{H_\lambda}^2(c) = \alpha_H^2(c).$$

By Sard's theorem, there exists a full Lebesgue measure set E of regular values λ of the function $\alpha_H^1 - \alpha_H^2$, which implies for $\lambda \in E$, 0 is a regular value of $\alpha_{H_\lambda}^1 - \alpha_{H_\lambda}^2(c)$, which implies the equation has only finitely many solutions on Γ, and at each solution, α^1 and α^2 has different derivatives when restricted to Γ. Note that this property is open in H, which implies the claim of our proposition is an open property. We only need to prove density.

Let $c_i(\lambda) \in \Gamma$, $i = 1, \cdots, N$ be the set on which $\alpha_{H_\lambda}^1 = \alpha_{H_\lambda}^2$; then since $\frac{d}{d\lambda}\alpha_{H_\lambda}^1(c_i) \neq \frac{d}{d\lambda}\alpha_{H_\lambda}^2(c_i)$, each $c_i(\lambda)$ is locally a monotone function in λ.

We now impose the assumption that H is a Kupka-Smale system; namely, all periodic orbits are non-degenerate, which holds on a residual set of H's (see [72]). In this setting, all invariant measures of rational rotation vectors are supported on hyperbolic periodic orbits, and by a further perturbation, we can ensure for each c there is only one minimal periodic orbit. Since the Aubry set is upper semi-continuous when there is only one static class, and the hyperbolic periodic orbit is structurally stable, we obtain that each hyperbolic periodic orbit is the Aubry set for an open set of c's. Let us denote by $R(H, \Gamma)$ the set of all $c \in \Gamma$ such that $\tilde{\mathcal{A}}_H^0(c)$ is a hyperbolic periodic orbit, then $R(H^1, \Gamma)$ and $R(H^2, \Gamma)$ are both open and dense in Γ. For each c_i, there is $d_i > 0$ and an open and dense subset of $\lambda \in B_{d_i}(0)$ such that $c_i(\lambda) \in R(H^1, \Gamma) \cap R(H^2, \Gamma)$. Let $d = \min d_i$, then there is an open and dense set of $\lambda \in B_d(0)$ such that for all $i = 1, \cdots, N$, $c_i(\lambda) \in R(\Gamma)$. For these c_i's, $\tilde{\mathcal{A}}_{H^1}^0(c_i), \tilde{\mathcal{A}}_{H^2}^0(c_i)$ are both hyperbolic periodic orbits. Using the Kupka-Smale theorem again, it is an open and dense property such that the stable and unstable manifolds of $\tilde{\mathcal{A}}_{H^1}^0(c_i)$ and $\tilde{\mathcal{A}}_{H^2}^0(c_i)$ intersect transversally. Due to dimension considerations, the intersection is a discrete set. Since each orbit in $\tilde{\mathcal{N}}_H(c_i) \setminus \tilde{\mathcal{A}}_H(c_i)$ is a heteroclinic orbit between $\tilde{\mathcal{A}}_{H^1}^0(c_i)$ and $\tilde{\mathcal{A}}_{H^2}^0(c_i)$, it is also a discrete set.

We have now proven the claim of our proposition holds for an arbitrarily small perturbation H_λ of H, and hence is dense on $\mathcal{V}_\sigma(H_*)$. □

An analogous statement holds for the asymmetric bifurcation case:

Proposition 8.20. *Let H_*, c_* be of asymmetric bifurcation type, then there is $\sigma > 0$ such that for an open and dense of $H \in \mathcal{V}_\sigma(H_*)$, such that there is a unique $c \in \overline{B_\sigma(c_*)} \cap \Gamma$ for which $\alpha^1_H(c) = \alpha^2_H(c)$. Moreover, at such (bifurcation) c, we have $\widetilde{\mathcal{A}}^0_{H^1}(c), \widetilde{\mathcal{A}}^0_{H^2}(c)$ both supported on hyperbolic periodic orbits. Moreover, $\widetilde{\mathcal{N}}^0_H(c) \setminus \widetilde{\mathcal{A}}^0_H(c)$ is a discrete set.*

Proof. The proof is nearly identical to Proposition 8.19 with the simplification that for σ small enough, $\widetilde{\mathcal{A}}_{H^2}(c)$ is always the same hyperbolic periodic orbit, and $\alpha_{H^2}(c)$ is a linear function. Since $\alpha_{H^1}(c)$ is convex on Γ, there is at most one bifurcation. The rest of the proof is identical.   □

Proof of Theorem 8.7. Case 1: Let H_*, c_* be of bifurcation Aubry-Mather type, and let $\sigma > 0$ be such that Proposition 8.19 holds on a open and dense subset $\mathcal{R}_1$ of $\mathcal{V}_\sigma(H_*)$. Then on $\Gamma \cap B_\sigma(c_*)$ there are at most finitely many bifurcations c_i, $i = 1, \cdots, N$. Moreover, for each bifurcation value c_i, there is $\sigma_i > 0$ such that for $c \in B_{\sigma_i}(c_i)$, the local Aubry sets are hyperbolic periodic orbits. This means that for each (non-bifurcation) $c \in \overline{B_{\sigma_i}(c_i)} \setminus c_i$, the sets $\widetilde{\mathcal{N}}^0_H(c) = \widetilde{\mathcal{A}}^0_H(c)$ are contractible (in fact finite). At the bifurcation values, the set $\widetilde{\mathcal{N}}^0_H(c_i) \setminus \widetilde{\mathcal{A}}^0_H(c_i)$ is discrete.

We now consider the set

$$\left(\Gamma \cap \overline{B_\sigma(c_*)} \right) \setminus \bigcup_{i=1}^{N} B_{\sigma_i}(c_i),$$

which is compact with finitely many connected components. On each of the components $\widetilde{\mathcal{A}}^0_H(c)$ has a unique static class. We apply Theorem 8.6 to get that there is $\sigma' > 0$ such that the dichotomy of Theorem 8.6 holds for a residual subset $\mathcal{R}_2$ of $H \in \mathcal{V}_{\sigma'}(H_*)$ on each of the connected components. The theorem follows by taking the intersection of $\mathcal{R}_1$ and $\mathcal{R}_2$ as well as the smallest value of σ and σ'.

Case 2: If (H_*, c_*) is of asymmetric bifurcation type, the proof is the same with Proposition 8.20 replacing Proposition 8.19.   □

Part III

Proving forcing equivalence

Chapter Nine

Aubry-Mather type at the single resonance

In this chapter we prove that for a single-resonance normal form system which satisfies the non-degeneracy conditions $[SR1_\lambda]$ or $[SR2_\lambda]$, every c in the resonance curve Γ is of either Aubry-Mather or bifurcation Aubry-Mather type. The main results are Theorems 9.3 and 9.5, which restate Propositions 3.9 and 3.10 in Part I.

At the end of this chapter we prove that the conditions hold on an open and dense set of Hamiltonians.

9.1   THE SINGLE MAXIMUM CASE

In this section we consider the Hamiltonian system

$$N_\epsilon = H_0(p) + \epsilon Z(\theta^s, p) + \epsilon R(\theta, p, t),$$

where $\theta = (\theta^s, \theta^f) \in \mathbb{T}^{n-1} \times \mathbb{T}$ and $p = (p^s, p^f) \in \mathbb{R}^{n-1} \times \mathbb{R}$. Consider the resonant curve

$$\Gamma = \{p \in B^n : \quad \partial_{p^s} H_0(p) = 0\} = \{p_*(p^f) = (p_*^s(p^f), p^f) : \quad p^f \in [a_-, a_+]\}.$$

We first consider the case $p \in \Gamma$ satisfy the condition $[SR1_\lambda]$, namely for each $p \in B_\lambda(p_0) \cap \Gamma$, $Z(\theta^s, p_*(p^f))$ has a unique global maximum at $\theta_*^s(p^f)$, and

$$D^{-1}I \le \partial_{pp}^2 H_0 \le DI, \quad \|Z\|_{C^3} \le 1, \quad \lambda I \le -\partial_{\theta^s\theta^s}^2 Z(\theta_*^s(p^f), p_*(p^f)) \le \lambda I,$$
$$\tag{9.1}$$

and that for $K > 0$, and $p_0 = p_*(a_0) \in \Gamma$,

$$\|R\|_{C_I^2(\mathbb{T}^n \times B_{K\sqrt{\epsilon}}(p_0) \times \mathbb{T})} \le \delta.$$

Note that we are using the rescaled norm C_I^2 (see (3.2)). We then set $K_1 = K/D$ and consider the local segment

$$\Gamma(\epsilon, p_0) = \{p_*(p^f) : \quad p^f \in [a_0 - K_1\sqrt{\epsilon}, a_0 + K_1\sqrt{\epsilon}]\} \subset \Gamma \cap B_{K\sqrt{\epsilon}}(p_0),$$

which is contained in $B_\lambda(p_0) \cap \Gamma$ if ϵ_0 is small enough depending on K. Throughout this section, we write $f = O(g)$ if $|f| \le C|g|$ for $C > 0$ that may depend only on D, λ and n.

Theorem 9.1 (See Sections 9.4 and 9.5, see also Theorem 3.1 of [13]). *Suppose that the Hamiltonian $N_\epsilon = H_0 + \epsilon Z + \epsilon R$ satisfies (9.1), and that $K_1 > 2$. There are $\delta_0, \epsilon_0 > 0$ and $C = C(D, \lambda, n) > 1$, such that if $0 < \delta < \delta_0$ and $0 < \epsilon < \max\{\epsilon_0, \sqrt{\delta}\}$, there is a C^2 map*

$$(\Theta^s, P^s)(\theta^f, p^f, t) : \mathbb{T} \times [a_0 - K_1\sqrt{\epsilon}/2, a_0 + K_1\sqrt{\epsilon}/2] \times \mathbb{T} \to \mathbb{T}^{n-1} \times \mathbb{R}^{n-1},$$

such that $\mathcal{C} = \{(\theta^s, p^s) = (\Theta^s, P^s)(\theta^f, p^f, t)\}$ is weakly invariant in the sense that the vector field is tangent to $\mathcal{C}$. $\mathcal{C}$ is contained in the set

$$V = \{(\theta, p, t) : \quad \|\theta^s - \theta^s_*(p^f)\| \le C^{-1}, \quad \|p^s - p^s_*(p^f)\| \le C^{-1}\}$$

and it contains all the invariant sets contained in V. Moreover, we have

$$\left\|\Theta^s(\theta^f, p^f, t) - \theta^s_*(p^f)\right\| \le C\delta, \quad \left\|P^s(\theta^f, p^f, t) - p^s_*(p^f)\right\| \le C\delta\sqrt{\epsilon},$$

$$\|\partial_{p^f}\Theta^s\| \le C\sqrt{\delta/\epsilon}, \quad \|\partial_{(\theta^f, t)}\Theta^s\| \le C\sqrt{\epsilon}, \quad \|\partial_{p^f}P^s\| \le C, \quad \|\partial_{(\theta^f, t)}P^s\| \le C\sqrt{\epsilon}.$$

The cylinder $\mathcal{C}$ is normally hyperbolic with its stable/unstable bundle projecting onto the θ^s direction.

Theorem 9.2 (See Section 9.6, see also Theorems 4.1 and 4.2 of [13]). *There are $\delta_0 = \delta_0(\lambda, n, D) > 0$ and $\epsilon_0 = \epsilon_0(\lambda, n, D, \delta)$ such that if $0 < \delta < \delta_0$ and $0 < \epsilon < \epsilon_0$, the Mañé set of the cohomology c satisfies*

$$\widetilde{\mathcal{N}}_{N_\epsilon}(c) \subset B_{\delta^{1/5}}(\theta^s_*) \times \mathbb{T} \times B_{\sqrt{\epsilon} \cdot \delta^{1/16}} \times \mathbb{T} \subset \mathbb{T}^{n-1} \times \mathbb{T} \times \mathbb{R}^n \times \mathbb{T}.$$

If u is a weak KAM solution of N_ϵ at c, then the set $\widetilde{\mathcal{I}}(u, c) \subset \mathbb{T}^n \times \mathbb{R}^n$ is contained in a $18\sqrt{D\epsilon}$-Lipschitz graph above $\mathbb{T}^n$.

The theorems as stated are analogous to the cited theorems in [13]. The main difference is that we now assume the much weaker assumption $\|R\|_{C^2_I} \le \delta$. Nevertheless, we now check that the method in [13] applies in the same way and leads to the estimates as stated.

9.2 AUBRY-MATHER TYPE AT SINGLE RESONANCE

Theorem 9.3. *Let $c = p_*(p^f)$ with $p^f \in [a_0 - K_1\sqrt{\epsilon}/4, a_0 + K_1\sqrt{\epsilon}/4]$ and there are $\epsilon_0, \delta_0 > 0$ such that for $N_\epsilon = H_0 + \epsilon Z + \epsilon R$ with $0 < \epsilon < \epsilon_0$ and $\delta < \delta_0$, then (N_ϵ, c) is of Aubry-Mather type with the hyperbolic cylinder given by the embedding*

$$\chi(\theta^f, p^f) = (\Theta^s, P^s)(\theta^f, p^f, 0),$$

where Θ^s, P^s are from Theorem 9.1.

We prove Theorem 9.3 assuming Theorems 9.1 and 9.2.

Proof of Theorem 9.3. Let us denote $\mathcal{C}$ the cylinder in Theorem 9.1 and $\mathcal{C}^0 = \mathcal{C} \cap \{t = 0\}$.

Let δ_0, ϵ_0 be small enough such that 9.2 applies for $0 < \epsilon < \epsilon_0$ and $0 < \delta < \delta_0$. In particular, these statements hold on a open set N_ϵ and c, as required by Definition 8.1. The embedding as described is a weakly normally hyperbolic invariant cylinder. Moreover, Corollary 7.7 implies that for any two tangent vectors $v, v' \in T_z \mathcal{C}^0$, we have $|d\Theta^s \wedge dP^s(v, v')| \leq C\sqrt{\delta}|d\theta^f \wedge dp^f(v, v')|$, and therefore for δ small enough

$$\left|(d\Theta^s \wedge dP^s + d\theta^f \wedge dp^f)(v, v')\right| \geq (1 - C\sqrt{\delta}) \geq \frac{1}{2}|d\theta^f \wedge dp^f(v, v')|.$$

Since the form $d\theta^f \wedge dp^f$ is non-degenerate on $\mathcal{C}^0$, the symplectic form $d\theta^s \wedge dp^f + d\theta^f \wedge dp^f$ is non-degenerate when restricted to $\mathcal{C}^0$.

Theorem 9.2 implies that for δ small enough, the Mañé set $\widetilde{\mathcal{N}}_{N_\epsilon}(c)$ is contained in the neighborhood V described in Theorem 9.1. Since the Mañé set is invariant, it must be contained in the maximally invariant cylinder $\mathcal{C}$. Therefore $\widetilde{\mathcal{A}}^0_{N_\epsilon}(c) \subset \widetilde{\mathcal{N}}^0_{N_\epsilon}(c) \subset \mathcal{C}^0$.

We now show that $\widetilde{\mathcal{A}}^0_{N_\epsilon}$ is a Lipschitz graph over θ^f. Let $(\theta_1, p_1), (\theta_2, p_2) \in \widetilde{\mathcal{A}}^0_{N_\epsilon}(c)$, then by Theorem 9.2 we have

$$\|p_2 - p_1\| \leq 18D\sqrt{\epsilon}\|\theta_2 - \theta_1\| \leq 18D\sqrt{\epsilon}\left(\|\theta_2^s - \theta_1^s\| + \|\theta_2^f - \theta_1^f\|\right).$$

By Theorem 9.1,

$$\|\theta_2^s - \theta_1^s\| \leq C(1 + \sqrt{\delta/\epsilon})\left(\|\theta_2^f - \theta_1^f\| + \|p_2 - p_1\|\right). \tag{9.2}$$

Combining everything, we get for some constant C_1 depending on n, D,

$$\left(1 - C_1(\sqrt{\epsilon} + \sqrt{\delta})\right)\|p_2 - p_1\| \leq C_1(\sqrt{\epsilon} + \sqrt{\delta})\|\theta_2^f - \theta_1^f\|,$$

which implies $\|p_2 - p_1\| \leq \|\theta_2^f - \theta_1^f\|$ if δ, ϵ are small enough depending only on C_1. Combining with (9.2), we get $\|(\theta_2, p_2) - (\theta_1, p_1)\| \leq 2\|\theta_2^f - \theta_1^f\|$ if ϵ, δ are small enough.

Finally, let u be a weak KAM solution for N_ϵ at cohomology c, and assume that $\widetilde{\mathcal{A}}^0_{N_\epsilon}(c)$ projects onto the θ^f component. Since the strong unstable manifold depends C^1 on the base point, the unstable manifold $W^u(\widetilde{\mathcal{A}}^0_H(c))$ is a Lipschitz manifold. Moreover, since the strong unstable direction projects onto the θ^s direction, $W^u(\widetilde{\mathcal{A}}^0_H(c))$ is locally a graph over $\theta = (\theta^f, \theta^s)$. This verifies 2b of Definition 8.1. We have verified all conditions in Definition 8.1 and therefore N_ϵ, c is of Aubry-Mather type. $\qquad\square$

9.3   BIFURCATIONS IN THE DOUBLE MAXIMA CASE

In this section we assume that for the Hamiltonian

$$N_\epsilon = H_0 + \epsilon Z + \epsilon R,$$

where Z satisfies the condition $[SR2_\lambda]$, namely for all $p \in B_\lambda(p_0) \cap \Gamma$, there exist two local maxima $\theta_1^s(p)$ and $\theta_2^s(p)$ of the function $Z(.,p)$ in $\mathbb{T}^{n-1}$ satisfying

$$\partial_{\theta^s}^2 Z(\theta_1^s(p), p) < \lambda I \quad , \quad \partial_{\theta^s}^2 Z(\theta_2^s(p), p) < \lambda I,$$

$$Z(\theta^s, p) < \max\{Z(\theta_1^f(p), p), Z(\theta_2^f(p), p)\} - \lambda \big(\min\{d(\theta^s - \theta_1^s), d(\theta^s - \theta_2^s)\} \big)^2.$$

Let $f : \mathbb{T}^{n-1} \to \mathbb{R}$ be a bump function satisfying the following conditions:

$$f|_{B_\lambda(\theta_1^s(p_0))} = 0, \quad f|_{B_\lambda(\theta_2^s(p_0))} = 1,$$

and $0 \le f \le 1$ otherwise. Define

$$Z_1 = Z - f, \quad Z_2 = Z - (1 - f);$$

then $Z_1(\cdot, p_0)$ has a unique maximum at $\theta_1^s(p_0)$, while $Z_2(\cdot, p_0)$ has a unique maximum at $\theta_2^s(p_0)$. There is $\kappa > 0$ depending only on λ such that the same holds for $p \in B_\kappa(p_0)$. To the Hamiltonian

$$N_\epsilon^i = N + \epsilon Z_i + \epsilon R, \quad i = 1, 2$$

we may apply Theorems 9.1 and 9.2 to obtain the existence of the NHIC $\mathcal{C}_i$ which contains the local Aubry sets $\mathcal{A}^i(c)$ for $c \in B_{\kappa/2}(p_0)$. Moreover, we may define the local alpha functions $\alpha^i(c) = \alpha_{N_\epsilon^i}(c)$ similar to Section 8.4. The cohomology $c \in \Gamma \cap B_{\kappa/2}(p_0)$ is of bifurcation type if and only if $\alpha^1(c) = \alpha^2(c)$, and is of Aubry-Mather type if and only if $\alpha^1(c) \ne \alpha^2(c)$.

Moreover, in this case we have the following analog of Theorem 9.2.

Theorem 9.4 (See Section 9.6, see also Theorem 4.5 of [13]). *Suppose Z satisfies condition $[SR2_\lambda]$ at $p = c$. Then there exists $\delta_0, \epsilon_0, \kappa > 0$ depending only on λ, n, D such that if $0 < \epsilon < \epsilon_0$ and $0 < \delta < \delta_0$, for every $c \in B_{\kappa/2}(p_0) \cap \Gamma$ the Aubry set satisfies*

$$\tilde{\mathcal{A}}(c) \subset \big(B(\theta_1^s, \delta^{1/5}) \cup B(\theta_2^s, \delta^{1/5}) \big) \times \mathbb{T} \times B(c, \sqrt{\epsilon} \delta^{1/16}) \times \mathbb{T}$$

$$\subset \mathbb{T}^{n-1} \times \mathbb{T} \times \mathbb{R}^n \times \mathbb{T}.$$

If, moreover, the projection $\theta^s(\mathcal{A}^0(c)) \subset \mathbb{T}^{n-1}$ is contained in one of the (disjoint) balls $B(\theta_i^s, \delta^{1/5})$, then the projection $\theta^s(\mathcal{N}^0(c)) \subset \mathbb{T}^{n-1}$ of the Mañé set is contained in the same ball $B(\theta_i^s, \delta^{1/5})$.

Theorem 9.5. *Suppose Z satisfies condition $[SR2_\lambda]$ at $p = c_*$. Then there* *exists* $\epsilon_0, \delta_0 > 0$, *such that if* $0 < \epsilon < \epsilon_0$, $0 < \delta < \delta_0$, *and* $\|R\|_{C_I^2} < \delta$, $N_\epsilon = H_0 + \epsilon Z + \epsilon R$ *is of bifurcation Aubry-Mather type.*

Proof. Let $V_1 = B_\lambda(\theta_1^s(p_0)) \times \mathbb{T} \subset \mathbb{T}^{n-1} \times \mathbb{T}$, and $V_2 = B_\lambda(\theta_2^s(p_0)) \times \mathbb{T}$, and define the local Hamiltonians N_ϵ^1 and N_ϵ^2 as in Definition 8.3. Theorem 9.4 implies the Aubry set $\tilde{\mathcal{A}}_{N_\epsilon}(c) = \tilde{\mathcal{A}}_{N_\epsilon^1}(c) \cap \tilde{\mathcal{A}}_{N_\epsilon^2}(c)$. We check that all the conditions of Definition 8.3 are satisfied. $\square$

9.4   HYPERBOLIC COORDINATES

The Hamiltonian flow admits the following equation of motion:

$$
\begin{cases}
\dot{\theta}^s = \partial_{p^s} H_0 + \epsilon \partial_{p^s} Z + \epsilon \partial_{p^s} R \\
\dot{p}^s = -\epsilon \partial_{\theta^s} Z - \epsilon \partial_{\theta^s} R \\
\dot{\theta}^f = \partial_{p^f} H_0 + \epsilon \partial_{p^f} Z + \epsilon \partial_{p^f} R \\
\dot{p}^f = -\epsilon \partial_{\theta^f} R \\
\dot{t} = 1
\end{cases}
\tag{9.3}
$$

The system (9.3) is a perturbation of

$$
\dot{\theta}^s = \partial_{p^s} H_0, \quad \dot{p}^s = -\epsilon \partial_{\theta^s} Z, \quad \dot{\theta}^f = \partial_{p^f} H_0, \quad \dot{p}^f = 0, \quad \dot{t} = 1,
$$

which admits a normally hyperbolic invariant cylinder

$$
\{(\theta_*^s(p^f), \theta^f, p_*^s(p^f), p^f, t): \quad \theta^f, t \in \mathbb{T}, \}.
$$

Set

$$
B(p^f) := \partial_{p^s p^s}^2 H_0(p_*(p^f)), \quad A(p^f) := -\partial_{\theta^s \theta^s}^2 Z(\theta_*^s(p^f), p_*(p^f));
$$

then as in [13], there is a positive definite matrix $T(p^f)$ such that

$$
\Lambda(p^f) = T(p^f) A(p^f) T(p^f) = T^{-1}(p^f) B(p^f) T^{-1}(p^f).
$$

We lift the equation to the universal cover, and consider the change of variable

$$
\begin{aligned}
x &= T^{-1}(p^f)(\theta^s - \theta_*^s(p^f)) + \epsilon^{-1/2} T(p^f)(p^s - p_*^s(p^f)) \\
y &= T^{-1}(p^f)(\theta^s - \theta_*^s(p^f)) - \epsilon^{-1/2} T(p^f)(p^s - p_*^s(p^f)), \\
I &= \epsilon^{-1/2}(p^f - a_0), \quad \Theta = \gamma \theta^f,
\end{aligned}
\tag{9.4}
$$

where $0 < \gamma < 1$ is a parameter to be determined.

Lemma 9.6 (Lemmas 3.1 and 3.2 of [13]). *For each* $p^f \in [a_-, a_+]$, *we have*

$\Lambda(p^f) \geq \sqrt{\lambda/D}I$. *We also have the following estimates:*

$$\|T\|_{C^2}, \|T^{-1}\|_{C^2}, \|\partial_{p^f}\theta_*^s\|_{C^0}, \|p_*^s\|_{C^2} = O(1),$$

$$\|\theta^s - \theta_*^s\|_{C^0} = O(\rho), \quad \|p^s - p_*^s\|_{C^0} = O(\sqrt{\epsilon}\rho),$$

where $\rho = \max\{\|x\|, \|y\|\}$.

Note we are using the regular C^2 norm in Lemma 9.6 since T depends only on H_0 and Z which are bounded in the regular norm.

Lemma 9.7. *The equation of motion in the new variables takes the form*

$$\dot{x} = \sqrt{\epsilon}\Lambda(a_0 + \sqrt{\epsilon}I)x + \sqrt{\epsilon}O(\delta + \rho^2),$$
$$\dot{y} = -\sqrt{\epsilon}\Lambda(a_0 + \sqrt{\epsilon}I)y + \sqrt{\epsilon}O(\delta + \rho^2),$$

and $\dot{I} = O(\sqrt{\epsilon}\delta)$.

Proof. The last equation is straightforward. We only prove the equation for $\dot{x}$ as the calculations for $\dot{y}$ are the same. In the original coordinates, we have

$$\dot{\theta}^s = B(p^f)(p^s - p_*^s(p^f)) + O(\|p^s - p_*^s(p^f)\|^2) + O(\delta\sqrt{\epsilon}),$$
$$\dot{p}^s = \epsilon A(p^f)(\theta^s - \theta_*^s(p^f)) + O(\epsilon\|\theta^s - \theta_*^s(p^f)\|^2) + O(\epsilon\delta),$$

where we used $\|\partial_p R\| = \epsilon^{-\frac{1}{2}}\|\partial_I R\| = O(\epsilon^{-\frac{1}{2}}\delta)$. Differentiating (9.4), using $\dot{p}^f = O(\epsilon\delta)$ and Lemma 9.6, we get

$$\dot{x} = T^{-1}\dot{\theta}^s + \epsilon^{-\frac{1}{2}}T\dot{p}^s + O(\sqrt{\epsilon}\delta)$$
$$= \sqrt{\epsilon}T^{-1}BT^{-1} \cdot \epsilon^{-\frac{1}{2}}T(p^s - p_*^s(p^f)) + \sqrt{\epsilon}TAT \cdot T^{-1}(\theta^s - \theta_*^s)$$
$$\quad + O(\sqrt{\epsilon}\delta + \sqrt{\epsilon}\rho^2)$$
$$= \sqrt{\epsilon}\Lambda(p^f)x + \sqrt{\epsilon}O(\delta + \rho^2).$$

$\square$

Lemma 9.8. *Suppose $\sqrt{\epsilon} \leq \delta$; then in the new coordinates (x, y, Θ, I, t) the linearized system is given by*

$$L = \begin{bmatrix} \sqrt{\epsilon}\Lambda & 0 & 0 & 0 & 0 \\ 0 & -\sqrt{\epsilon}\Lambda & 0 & 0 & 0 \\ 0 & 0 & 0 & 0 & 0 \\ 0 & 0 & 0 & 0 & 0 \\ 0 & 0 & 0 & 0 & 0 \end{bmatrix} + \sqrt{\epsilon}O(\delta\gamma^{-1} + \rho + \gamma),$$

where $\rho = \max\{\|x\|, \|y\|\}$.

Proof. In the original coordinates, the linearized equation is given by the matrix

$$\tilde{L} = \begin{bmatrix} O(\sqrt{\epsilon}\delta) & B + O(\delta) & O(\sqrt{\epsilon}\delta) & \partial^2_{p^f p^s} H_0 + O(\delta) & 0 \\ \epsilon A + O(\epsilon\rho) & O(\sqrt{\epsilon}\delta) & 0 & O(\sqrt{\epsilon}\delta) & 0 \\ O(\sqrt{\epsilon}\delta) & O(1) & O(\sqrt{\epsilon}\delta) & O(1) & 0 \\ 0 & O(\sqrt{\epsilon}\delta) & 0 & O(\sqrt{\epsilon}\delta) & 0 \\ 0 & 0 & 0 & 0 & 0 \end{bmatrix} + O(\epsilon\delta).$$

The coordinate change matrix is

$$\left[\frac{\partial(\theta^s, p^s, \theta^f, p^f, t)}{\partial(x, y, \Theta, I, t)}\right] = \begin{bmatrix} T/2 & T/2 & 0 & O(\sqrt{\epsilon}) & 0 \\ \sqrt{\epsilon}T^{-1}/2 & -\sqrt{\epsilon}T^{-1}/2 & 0 & \sqrt{\epsilon}\partial_{p^f} p^s_* + O(\epsilon\rho) & 0 \\ 0 & 0 & \gamma^{-1} & 0 & 0 \\ 0 & 0 & 0 & \sqrt{\epsilon} & 0 \\ 0 & 0 & 0 & 0 & 1 \end{bmatrix}.$$

The product is

$$\tilde{L}\left[\frac{\partial(\theta^s, p^s, \theta^f, p^f, t)}{\partial(x, y, \Theta, I, t)}\right] = O(\epsilon\delta\gamma^{-1} + \epsilon\rho) +$$
$$\begin{bmatrix} \sqrt{\epsilon}BT^{-1}/2 + O(\sqrt{\epsilon}\delta) & -\sqrt{\epsilon}BT^{-1}/2 + O(\sqrt{\epsilon}\delta) & O(\sqrt{\epsilon}\delta\gamma^{-1}) & O(\sqrt{\epsilon}\delta) & 0 \\ \epsilon AT/2 & \epsilon AT/2 & 0 & 0 & 0 \\ O(\sqrt{\epsilon}) & O(\sqrt{\epsilon}) & O(\sqrt{\epsilon}\delta\gamma^{-1}) & O(\sqrt{\epsilon}) & 0 \\ 0 & 0 & 0 & 0 & 0 \\ 0 & 0 & 0 & 0 & 0 \end{bmatrix}.$$

Most of the computations are straightforward, with the exception of the fourth row, first column, which contains the following cancellation:

$$\partial^2_{p^s p^f} H_0 \partial_{p^f} p^s_* + \partial^2_{p^f p^f} H_0 = \partial_{p^f}\left(\partial_{p^f} H_0(p_*(p^f))\right) = 0,$$

since $\partial_{p^f} H_0(p_*(p^f)) = 0$ for every p^f by definition.

The differential of the inverse coordinate change is

$$\left[\frac{\partial(x, y, \Theta, I, t)}{\partial(\theta^s, p^s, \theta^f, p^f, t)}\right] = \begin{bmatrix} T^{-1} & \epsilon^{-1/2}T & 0 & O(\epsilon^{-1/2}\lambda^{-1/4}) & 0 \\ T^{-1} & -\epsilon^{-1/2}T & 0 & O(\epsilon^{-1/2}\lambda^{-1/4}) & 0 \\ 0 & 0 & \gamma & 0 & 0 \\ 0 & 0 & 0 & \epsilon^{-1/2} & 0 \\ 0 & 0 & 0 & 0 & 1 \end{bmatrix}.$$

Finally, the new matrix of the linearized equation is

$$L = \left[\frac{\partial(x, y, \Theta, I, t)}{\partial(\theta^s, p^s, \theta^f, p^f, t)}\right] \tilde{L} \left[\frac{\partial(\theta^s, p^s, \theta^f, p^f, t)}{\partial(x, y, \Theta, I, t)}\right]$$

$$= O(\sqrt{\epsilon}(\delta\gamma^{-1} + \rho)) + \begin{bmatrix} \sqrt{\epsilon}\Lambda & 0 & 0 & 0 & 0 \\ 0 & -\sqrt{\epsilon}\Lambda & 0 & 0 & 0 \\ O(\sqrt{\epsilon}\gamma) & O(\sqrt{\epsilon}\gamma) & O(\sqrt{\epsilon}\gamma) & 0 & 0 \\ 0 & 0 & 0 & 0 & 0 \\ 0 & 0 & 0 & 0 & 0 \end{bmatrix}.$$

□

9.5 NORMALLY HYPERBOLIC INVARIANT CYLINDER

We state the following abstract statement for existence of normally hyperbolic invariant manifolds, given in [13].

Let $F : \mathbb{R}^n \to \mathbb{R}^n$ be a C^1 vector field. We split the space $\mathbb{R}^n$ as $\mathbb{R}^{n_u} \times \mathbb{R}^{n_s} \times \mathbb{R}^{n_c}$, and denote by $z = (u, s, c)$ the points of $\mathbb{R}^n$. We denote by (F_u, F_s, F_c) the components of F:

$$F(x) = (F_u(z), F_s(z), F_c(z)).$$

We study the flow of F in the domain

$$\Omega = B^u \times B^s \times \Omega^c,$$

where B^u and B^s are the open Euclidean balls of radius r_u and r_s in $\mathbb{R}^{n_u}$ and $\mathbb{R}^{n_s}$, and $\Omega^c = \Omega^{c_1} \times \mathbb{R}^{c_2}$ is a convex open subset of $\mathbb{R}^{n_c}$. We denote by

$$L(z) = dF(z) = \begin{bmatrix} L_{uu}(z) & L_{us}(z) & L_{uc}(z) \\ L_{su}(z) & L_{ss}(z) & L_{sc}(z) \\ L_{cu}(z) & L_{cs}(z) & L_{cc}(z) \end{bmatrix}$$

the linearized vector field at point z. We assume that $\|L(z)\|$ is bounded on Ω, which implies that each trajectory of F is defined until it leaves Ω. We denote by $W^c(F, \omega)$ the union of all full orbits contained in Ω, $W^{sc}(F, \Omega)$ the set of points whose positive orbit remains inside Ω, and by $W^{uc}(F, \Omega)$ the set of points whose negative orbit remains inside Ω.

Let us further consider a positive parameter $b > 0$, and consider the set $\Omega_b^{c_2} = B_b(\Omega^{c_2})$ and $\Omega_b^c = \Omega^{c_1} \times \Omega_b^{c_2}$.

Proposition 9.9 (Proposition A.6 of [13]). *Let $F : \mathbb{R}^{n_u} \times \mathbb{R}^{n_s} \times \Omega_b^c \to \mathbb{R}^{n_u} \times \mathbb{R}^{n_s} \times \mathbb{R}^{n_c}$ be a C^2 vector field. Assume that there exist $\alpha, m, \sigma > 0$ such that*

- $F_u(u, s, c) \cdot u > 0$ on $\partial B^u \times \bar{B}^s \times \bar{\Omega}_b^c$.
- $F_s(u, s, c) \cdot s < 0$ on $\bar{B}^u \times \partial B^s \times \bar{\Omega}_b^c$.

- $L_{uu}(z) \geqslant \alpha I$, $L_{ss}(z) \leqslant -\alpha I$ for each $x \in \Omega_b$ in the sense of quadratic forms.
- $\|L_{us}\| + \|L_{uc}\| + \|L_{ss}\| + \|L_{sc}\| + \|L_{cu}\| + \|L_{cs}\| + \|L_{cc}\| \leqslant m$ evaluated at each $z \in \Omega_b$.
- $\|L_{us}\| + \|L_{uc}\| + \|L_{ss}\| + \|L_{sc}\| + \|L_{cu}\| + \|L_{cs}\| + \|L_{cc}\| + 2\|F_{c_2}\|/b \leqslant m$ evaluated at each $z \in \Omega_b - \Omega$.

Assume furthermore that

$$K := \frac{m}{\alpha - 2m} \leqslant \frac{1}{8},$$

then there exist C^2 maps

$$w^{sc} : B^s \times \Omega_b^c \to B^u, \quad w^{uc} : B^u \times \Omega_b^c \to B^s, \quad w^c : \Omega_b^c \to B^u \times B^s$$

satisfying the estimates

$$\|dw^{sc}\| \leqslant K, \quad \|dw^{uc}\| \leqslant K, \quad \|dw^c\| \leqslant 2K,$$

the graphs of which respectively contain $W^{sc}(F, \Omega), W^{uc}(F, \Omega), W^c(F, \Omega)$. Moreover, the graphs of the restrictions of w^{sc}, w^{uc}, and w^c to, respectively, $B^s \times \Omega^c$, $B^u \times \Omega^c$, and Ω^c, are tangent to the flow.

There exists an invariant C^1 foliation of the graph of w^{uc} whose leaves are graphs of K-Lipschitz maps above B^u. The set $W^{uc}(F, \Omega)$ is a union of leaves: it has the structure of an invariant C^1 lamination. Two points x, x' belong to the same leaf of this lamination if and only if $d(x(t), x'(t))e^{t\alpha/4}$ is bounded on $\mathbb{R}^-$.

If in addition there exists a group G of translations of $\mathbb{R}^{n_{c_1}}$ such that $F \circ (id \otimes id \otimes g \otimes id) = F$ for each $g \in G$, then the maps w^* can be chosen such that

$$w^{sc} \circ (id \otimes g \otimes id) = w^{sc}, \quad w^{uc} \circ (id \otimes g \otimes id) = w^{uc}, \quad w^c \circ (g \otimes id) = w^c \quad (9.5)$$

for each $g \in G$. The lamination is also translation invariant.

Proof of Theorem 9.1. Consider the equation of N_ϵ in the (x, y, θ^f, p^f, t) coordinates, and set $B^u = \{\|x\| < \rho\}$, $B^s = \{\|y\| < \rho\}$, $\Omega^{c_2} = \{\|I^f\| \leq K_1/2\}$, $\Omega^{c_1} = \{(\theta^f, t)\}$ with the group translation $\mathbb{Z}^2$, $\alpha = \sqrt{\lambda/4D}$, $b = 1 < K_1/2$. Note that our choice of b implies $\Omega_b^{c_2} \subset \{\|I^f\| < K_1\}$. We fix $\gamma = \sqrt{\delta}$

According to Lemmas 9.6 and 9.7, we have for $\|x\| = \rho$,

$$\dot{x} \cdot x \geq \sqrt{\epsilon}\alpha\rho^2 - \rho\sqrt{\epsilon}O(\delta + \rho^2) = \sqrt{\epsilon}\rho\left(\rho - O(\delta + \rho^2)\right) > 0$$

as long as $\rho \geq C\delta$ for C large enough. This verifies the first bullet point assumption. The second assumption is verified in the same way.

By Lemma 9.8, we have for $C_1 > 1$ depending on λ, D, n,

$$\|L_{us}(z)\| + \|L_{uc}(z)\| + \|L_{ss}(z)\| + \|L_{sc}(z)\| + \|L_{cu}(z)\| + \|L_{cs}(z)\| + \|L_{cc}(z)\|$$
$$= \sqrt{\epsilon}O(\delta\gamma^{-1} + \rho + \gamma) \leq C_1\sqrt{\epsilon}(\sqrt{\delta} + \rho) =: m/2,$$

and $L_{uu} \geq (2\alpha - m)I \geq \alpha I$ as long as $m < \alpha$. This is possible as long as $\epsilon, \delta, \rho < C_1^{-1} < \alpha$. Finally, we check that if $z \in \Omega_b \setminus \Omega$, $\|F_{c_2}(z)\| = O(\sqrt{\epsilon}\delta) \leq C_1\sqrt{\epsilon}(\sqrt{\delta} + \rho)/2 = m/4$ if C_1 is chosen large enough. These estimates ensure all the bullet point conditions are satisfied. By choosing C_1 even smaller, we can ensure $m < \alpha/10$ which implies $K = m/(\alpha - 2m) \leq 1/8$.

To summarize, we have shown all conditions of Proposition 9.9 are satisfied if $0 < \epsilon < C_1^{-1}$, $0 < \delta < C_1^{-1}$, $C\delta \leq \rho \leq C_1^{-1}$. We apply the Proposition twice, once for $\rho = C_1^{-1}$ and once for $\rho = C\delta$. The first application shows the invariant cylinder is the maximal invariant set in the set $\Omega^c = \{\|x\|, \|y\| \leq C_1^{-1}\}$. The second application shows that the cylinder is in fact contained in the set $\{\|x\|, \|y\| \leq C\delta\}$. Moreover, in the second application, we get the estimate $m = O(\sqrt{\epsilon}\delta)$, which allows the estimate $K = O(\sqrt{\epsilon}\delta)$.

We now return to the original coordinates using the formula

$$\Theta^s(\theta^f, p^f, t) = \theta_*^s(p^f) + \frac{1}{2}T(p^f) \cdot (w_u^c + w_s^c)(\gamma\theta^f, a_0 + \epsilon^{-1/2}p^f, t)$$
$$P^s(\theta^f, p^f, t) = p_*^s(p^f) + \frac{\sqrt{\epsilon}}{2}T^{-1}(p^f) \cdot (w_u^c - w_s^c)(\gamma\theta^f, a_0 + \epsilon^{-1/2}p^f, t).$$
(9.6)

All the estimates stated in Theorem 9.1 follow directly from these expressions, and from the fact that $\{\|x\|, \|y\| \leq C\delta\}$, $\|dw^c\| \leqslant 2K = O(\sqrt{\epsilon}\delta)$. $\square$

9.6 LOCALIZATION OF THE AUBRY AND MAÑÉ SETS

Let $D_1 = 2D$, the Hamiltonian N_ϵ satisfies the following estimates if δ is small enough depending only on D:

$$D_1^{-1}I \leq \partial_{pp}^2 N_\epsilon \leq DI, \quad \|\partial_{\theta p}^2 N_\epsilon\| \leq 2\sqrt{\epsilon}, \quad \|\partial_{\theta\theta}^2 N_\epsilon\| \leq 3\epsilon.$$

Let $L(\theta, p, t)$ be the Lagrangian associated to N_ϵ, and L_0 the Lagrangian of H_0.

Lemma 9.10. *The following estimates hold for the Lagrangian L.*

1. *(Lemma 4.1 of [13]) For $K > 0$, the image of the set $\mathbb{T}^n \times B_{K\sqrt{\epsilon}} \times \mathbb{T}$ under the diffeomorphism $\partial_p N_\epsilon$ contains the set $\mathbb{T}^n \times B_{K_1\sqrt{\epsilon}}(c) \times \mathbb{T}$, where $K_1 = K/(4D_1)$.*
2. *(Lemma 4.2 of [13]) The estimate*

$$\|\partial_{\theta v}^2 L\|_{C^0} \leq 2D_1\sqrt{\epsilon}, \quad \|\partial_{\theta\theta}^2 L\|_{C^0} \leq 3\epsilon$$

holds on $\mathbb{T}^n \times B_{K_1\sqrt{\epsilon}}(c) \times \mathbb{T}$.

3. *(Lemma 4.3 of [13]) For* $v \in B_{K_1\sqrt{\epsilon}}(c)$, *we have*

$$|L(\theta, v, t) - (L_0(v) - \epsilon Z(\theta^s, c))| \leq 2\epsilon\delta.$$

4. *(Lemma 4.4 of [13]) The alpha function* $\alpha_{N_\epsilon}(c)$ *satisfies*

$$|\alpha_{N_\epsilon}(c) - (H_0(c) + \epsilon \max Z(\cdot, c))| \leq 2\epsilon\delta.$$

5. *(Lemma 4.5 of [13]) There is* $C > 1$ *depending only on* D *such that if* $\epsilon < C^{-1}\delta$, *we have the estimates*

$$L(\theta, v, t) - c \cdot v + \alpha(c) \geqslant \|v - \partial H_0(c)\|^2/(4D_1) - \epsilon\hat{Z}_c(\theta^s) - 4\epsilon\delta \qquad (9.7)$$

$$L(\theta, v, t) - c \cdot v + \alpha(c) \leqslant D_1\|v - \partial H_0(c)\|^2 - \epsilon\hat{Z}_c(\theta^s) + 4\epsilon\delta \qquad (9.8)$$

for each $(\theta, v, t) \in \mathbb{T}^n \times \mathbb{R}^n \times \mathbb{R}$, *where* $\hat{Z}_c(\theta^s) := Z(\theta^s, c) - \max_{\theta^s} Z(\theta^s, c)$.

Proof. We remark that in [13] it is assumed that $\|\partial^2_{\theta p} N_\epsilon\| \leq 2\epsilon$ instead of $2\sqrt{\epsilon}$ as we assumed. However the same calculations as given in the cited lemmas prove Lemma 9.10, once appropriate changes are made. □

Proposition 9.11. *For each* $c \in \mathbb{R}^n$, *the weak KAM solution* u *of cohomology* c *is* $3\sqrt{D_1\epsilon}/2$*-semi-concave.*

Proof of Theorem 9.2. The proofs of Theorem 4.1 and 4.2 in [13] rely on the estimates 3, 4, and 5 in Lemma 9.10, as well as Proposition 9.11. We have arrived at the same estimates using a weaker assumption, the only change is that we replaced the constant D with $D_1 = 2D$. The rest of the proofs are exactly identical. □

Proof of Theorem 9.4. This is similar to Theorem 9.2. The proof of Theorem 4.5, [13] applies once we take into account Lemma 9.10 and Proposition 9.11. □

9.7 GENERICITY OF THE SINGLE-RESONANCE CONDITIONS

We prove Proposition 3.2 in this section, which is a direct consequence of the Lemma 9.12. We say a local maximum x_0 of a function $g(x)$ is non-degenerate if $\partial^2_{xx} g(x_0)$ is strictly negative definite.

Lemma 9.12. *Suppose* $r \geq 4$. *Let* $a_1 < a_2$. *Let* $\mathcal{R}$ *be the set of* $f = f(x, a) \in C^r(\mathbb{T} \times [a_1, a_2])$ *such that* $f(\cdot, a)$ *has unique non-degenerate global maximum for all but finitely many* a's, *and that at those exceptional points there are two non-degenerate maxima. Then* $\mathcal{R}$ *is open and dense in* $C^r(\mathbb{T} \times [a_1, a_2])$.

Proof. Letting $[a_1, a_2] \subset (b_1, b_2)$, we consider the subspace $\mathcal{D}$ of the jet space $J^3(\mathbb{T} \times (b_1, b_2))$, defined by $\partial_x f(x, a) = \partial_{xx} f(x, a) = \partial_{xxx} f(x, a) = 0$, and let $\mathcal{D}_1$ be the restriction of $\mathcal{D}$ to base points in $\mathbb{T} \times [a_1, a_2]$. By Thom's transverality theorem (see for example Theorem 2.8 of [48]), if $r \geq 4$, the set $\mathcal{R}_1 \subset C^r(\mathbb{T} \times (b_1, b_2))$ such that the jet map $f \mapsto J^3 f$ is transversal to $\mathcal{D}$ at $\mathcal{D}_1$ is open and dense. Moreover, since $\mathcal{D}$ has co-dimension 3 and $\mathbb{T} \times (b_1, b_2)$ has dimension 2, for each $g \in \mathcal{R}_1$, $J^3 g$ is disjoint from $\mathcal{D}_1$, i.e., for all $a \in [a_1, a_2]$ such that $\partial_x f(x, a) = \partial_{xx} f(x, a) = 0$, we have $\partial_{xxx} f(x, a) \neq 0$. It follows that any degenerate critical point cannot be a local maximum. We have proven that if $g \in \mathcal{R}_1$, then $g(\cdot, a)$ has no degenerate local maxima for all $a \in [a_1, a_2]$.

Claim 1: Suppose $g \in \mathcal{R}_1$. Let $a_0 \in [a_1, a_2]$. Suppose $x_1, \ldots, x_k$ are the maxima of $g(\cdot, a_0)$. By the implicit function theorem, there exist $\epsilon, \delta > 0$, such that for each $g_1 \in \mathcal{V}_\delta(g)$ (the neighborhood is taken in $C^r(\mathbb{T} \times (b_1, b_2))$), there exist smooth maps $x_j^{g_1} : [a_0 - \epsilon, a_0 + \epsilon] \to \mathbb{T}$, $j = 1, \ldots, k$, such that $x_j^{g_1}$ is continuous in g_1 in C^3 topology, with $x_j^g(a_0) = x_j$. $x_j^{g_1}(a)$ are the local maxima of $g_1(\cdot, a)$, and

$$\max_x g_1(x, a) = \max_{1 \leq j \leq k} g_1(x_j^{g_1}(a), a), \quad \forall a \in [a_0 - \epsilon, a_0 + \epsilon]. \tag{9.9}$$

To prove Claim 1, note that there are at most finitely many global maxima of $g(\cdot, a_0)$, since a maximum is isolated (since it's non-degenerate) and cannot accumulate (the limit point is an non-isolated maximum). Since a limit point (in both a and in g) of maxima is a maximum, this implies for ϵ, δ sufficiently small, $a \in (a_0 - 2\epsilon, a_0 + 2\epsilon)$, and $g_1 \in \mathcal{V}_\delta(g)$, any maximum of $g_1(\cdot, a)$ must be in a neighborhood of $\{x_j : 1 \leq j \leq k\}$. Since $x_j^{g_1}$ is the unique local maximum in a neighborhood (again, due to non-degeneracy), we conclude that any maximum of $g_1(\cdot, a)$ must be taken among $x_j^{g_1}$, $1 \leq j \leq k$.

Claim 2: Let g, a_0, ϵ, δ be as in Claim 1. The following property holds for g_1 in an open and dense subset of $\mathcal{V}_\delta$: for each $a \in [a_0 - \epsilon, a_0 + \epsilon]$, $g_1(\cdot, a)$ has at most two maxima, each maximum is non-degenerate, and at each a for which there are two maxima, say $x_i^{g_1}(a)$ and $x_j^{g_1}(a)$, we have

$$\frac{d}{da} g_1(x_i^{g_1}(a), a) \neq \frac{d}{da} g_1(x_j^{g_1}(a), a).$$

As a result, there are at most finitely many points with two maxima. Before proving Claim 2, we note that our lemma follows from it by covering $[a_1, a_2]$ with finitely many intervals on which Claim 2 holds.

To prove Claim 2, let U_j, $1 \leq j \leq k$, be disjoint open sets, $x_j^{g_1} \in U_j$ for all $g_1 \in \mathcal{V}_\delta(g)$, ρ_j be compactly supported C^∞ functions such that $\rho_j|U_j = 1$, and that $\text{supp}\,\rho_j$ are mutually disjoint. Fix a $g_1 \in \mathcal{V}_\delta(g)$, $\beta = (\beta_1, \ldots, \beta_k)$,

$\gamma = (\gamma_1, \ldots, \gamma_k)$, define

$$G(x, a, \beta, \gamma) = g_1(x, a) + \sum_{j=1}^{k} \beta_j \rho_j + \sum_{j=1}^{k} \gamma_j a \rho_j,$$

and for $\sigma > 0$ small enough, (the evaluation map)

$$eG : [a_0 - \epsilon, a_0 + \epsilon] \times B_\sigma(0) \times B_\sigma(0) \to \mathbb{R}^{2k},$$

$$(eG(a, \beta, \gamma))_{2j-1,2j} = (G(x_j^{G(\cdot,\cdot,\beta,\gamma)}, a, \beta, \gamma), \frac{d}{da} G(x_j^{G(\cdot,\cdot,\beta,\gamma)}, a, \beta, \gamma)),$$

$$1 \le j \le k.$$

Note that $(eG(a, \beta))_{2j-1,2j} = (g_1(x_j^{g_1}(a), a) + \beta_j, \frac{d}{da} g_1(x_j^{g_1}(a), a) + \gamma_j)$, and hence the map is transversal to $\mathbb{R}^{2k}$. Define

$$\Delta_{2,2} = \{(y_j, v_j)_{j=1}^{k} \in \mathbb{R}^{2k} : \exists j_1 < j_2 \text{ such that } y_{j_1} = y_{j_2}, v_{j_1} = v_{j_2}\},$$

$$\Delta_3 = \{(y_j, v_j)_{j=1}^{k} \in \mathbb{R}^{2k} : \exists j_1 < j_2 < j_3 \text{ such that } y_{j_1} = y_{j_2} = y_{j_3}\}.$$

Since $\Delta_{2,2}$ and Δ_3 are both co-dimension 2, Sard's theorem implies that for a generic β, the image of $eG(\cdot, \beta)$ has no intersections with either set. This observation proves the density part of Claim 2. The condition is clearly open. This proves Claim 2 and our lemma.   □

Chapter Ten

Normally hyperbolic cylinders at double resonance

We prove Theorem 4.4 in this chapter. There are two cases:

1. (Simple critical homology) In this case we show the homoclinic orbit $\eta^0_{\pm h}$ can be extended to periodic orbits both in positive and negative energy. The union of these periodic orbits forms a C^1 normally hyperbolic invariant manifold (which is homotopic to a cylinder with a puncture).
2. (Non-simple homology) In this case we show that for positive energy, there exist periodic orbits shadowing $\eta^0_{h_1}$ and $\eta^0_{h_2}$ in a particular order.

Our strategy is to prove the existence of these periodic orbits as hyperbolic fixed points of composition of local and global maps. A main technical tool to prove the existence and uniqueness of these fixed points is the Conley-McGehee isolation block ([68]). The plan of this section is as follows.

In Section 10.1 we state a standard normal form near the hyperbolic fixed point.

In Section 10.2 we study the property of the Shil'nikov boundary value problem.

In Section 10.3, we apply results of Section 10.2 to establish strong hyperbolicity of the local map Φ^*_{loc} as well as the existence of unstable cones. Since the global maps Φ^*_{glob} are Poincaré maps of bounded transition time, they have bounded norms and the linearizations of the proper compositions $\Phi^*_{\mathrm{glob}}\Phi^*_{\mathrm{loc}}$ are dominated by the local component.

In Section 10.4, we define isolating blocks of Conley-McGehee [68] and use it to construct isolating blocks for the proper compositions of $\Phi^*_{\mathrm{glob}}\Phi^*_{\mathrm{loc}}$, assuming conditions $[DR1^c] - [DR4^c]$. We then apply the same analysis to $\Phi^*_{\mathrm{glob}}\Phi^*_{\mathrm{loc}}\cdots\Phi^*_{\mathrm{glob}}\Phi^*_{\mathrm{loc}}$ and construct families of shadowing orbits in the non-simple case.

In Section 10.5 we complete the proof of Theorem 4.4 by showing that periodic orbits constructed in the previous two sections form a normally hyperbolic invariant cylinder. Moreover, they coincide with the shortest geodesics for the Jacobi metric.

In Section 10.6 we prove Lemma 4.2 which describes, in the non-simple case, the order at which the simple periodic orbits are shadowed.

10.1 NORMAL FORM NEAR THE HYPERBOLIC FIXED POINT

We describe a normal form near the hyperbolic fixed point (assumed to be $(0,0)$) of the slow Hamiltonian $H^s : \mathbb{T}^2 \times \mathbb{R}^2 \to \mathbb{R}$. For the rest of this section, we drop the superscript s to abbreviate notations. In a neighborhood of the origin, there exists a symplectic linear change of coordinates under which the system has the normal form

$$H(u_1, u_2, s_1, s_2) = \lambda_1 s_1 u_1 + \lambda_2 s_2 u_2 + O_3(s, u).$$

Here $s = (s_1, s_2)$, $u = (u_1, u_2)$, and $O_n(s, u)$ stands for a function bounded by $C\|(s, u)\|^n$. By taking a standard straightening coordinate change, we get:

Lemma 10.1. *After an C^{r-1} symplectic coordinate change Φ, the Hamiltonian takes the form*

$$N = H \circ \Phi = \lambda_1 s_1 u_1 + \lambda_2 s_2 u_2 + \sum_{i,j=1,2} s_i u_j O_1(s, u),$$

and the equation is

$$\begin{bmatrix} \dot{s} \\ \dot{u} \end{bmatrix} = \begin{bmatrix} -\Lambda s + sO_1(s, u) \\ \Lambda u + uO_1(s, u) \end{bmatrix}, \tag{10.1}$$

where $\Lambda = diag\{\lambda_1, \lambda_2\}$.

Proof. Since $(0,0)$ is a hyperbolic fixed point, for sufficiently small $r > 0$, there exist stable manifold $W^s = \{(u = U(s), |s| \leq r\}$ and unstable manifold $W^u = \{s = S(u), |u| \leq r\}$ containing the origin. All points on W^s converge to $(0,0)$ exponentially in forward time, while all points on W^u converge to $(0,0)$ exponentially in backward time. In a Hamiltonian system, the unstable/stable manifolds of equilibria is Lagrangian. Indeed, if $z \in W^u$ and $v_1, v_2 \in T_z W^u$, then $\omega(v_1, v_2) = \omega(D\phi_H^{-t} v_1, D\phi_H^{-t} v_2) \to 0$ as $t \to \infty$. We now claim that the coordinate change

$$(s, u) \mapsto (s^+, u^+) = (s, u - U(s))$$

is symplectic. Indeed $u = U(s)$ is an Lagrangian graph, and hence $U(s) = \nabla G(s)$ for a smooth function G. Then $d(s^+) \wedge d(u^+) = ds \wedge du + (D^2 G(s) ds) \wedge ds = ds \wedge du$. Similarly, $(s^+, u^+) \mapsto (s', u') = (s^+ - S(u^+), u^+)$ is also symplectic. Under the new coordinates, $W^s = \{u' = 0\}$ and $W^u = \{s' = 0\}$. We abuse notation and keep using (s, u) to denote the new coordinate system.

Under the new coordinate system, the Hamiltonian has the form

$$H(s, u) = \lambda_1 s_1 u_1 + \lambda_2 s_2 u_2 + H_1(s, u),$$

where $H(s, u) = O_3(s, u)$ and $H_1(s, u)|_{s=0} = H_1(s, u)|_{u=0} = 0$. The Hamiltonian and the vector field take the desired form under these coordinates. $\square$

10.2   SHIL'NIKOV'S BOUNDARY VALUE PROBLEM

The local maps Φ_{loc} are Poincaré maps from a section $\Sigma_{\pm}^{s}$ transversal to s_1 axis, to $\Sigma_{\pm}^{u}$ transversal to the u_1 axis. The main tool for their study is the Shil'nikov boundary value problem.

Proposition 10.2 (Shil'nikov, Lemmas 2.1, 2.2, 2.3 of [73]). *There exist $C > 0$, $\delta_0 > 0$ and $T_0 > 0$ such that for each $0 < \delta < \delta_0$, any $s^{in} = (s_1^{in}, s_2^{in})$, $u^{out} = (u_1^{out}, u_2^{out})$ with $\|(s, u)\| \leq \delta$, and any large $T > T_0$, there exists a unique solution $(s^T, u^T) : [0, T] \to B_{C\delta}$ of the system (10.1) with the property $s^T(0) = s^{in}$ and $u^T(T) = u^{out}$. Moreover, we have*

$$\|s^T(t)\| \leq C\delta e^{-\lambda_1 t/2}, \quad \|u^T(t)\| \leq C\delta e^{-\lambda_1(T-t)/2},$$

and

$$\left\| \frac{\partial s^T(t)}{\partial s^{in}} \right\| + \left\| \frac{\partial u^T(t)}{\partial s^{in}} \right\| \leq C e^{-\lambda_1 t/2}, \quad \left\| \frac{\partial s^T(t)}{\partial u^{out}} \right\| + \left\| \frac{\partial u^T(t)}{\partial u^{out}} \right\| \leq C e^{-\lambda_1(T-t)/2}.$$

Corollary 10.3. *Let (s^T, u^T) be the solution in Proposition 10.2 with fixed boundary values s^{in}, u^{out}. Then*

$$\left\| \frac{d}{dT}(s^T(t), u^T(t)) \right\| \leq C e^{-\lambda_1(T-t)/2}, \quad \left\| \frac{d}{dT}(s^T(T-t), u^T(T-t)) \right\| \leq C e^{-\lambda_1 t/2},$$

$$\tag{10.2}$$

and

$$\frac{d}{d\tau}\bigg|_{\tau=T}(s^\tau(T), u^\tau(T)) = -X_N(s^T(T), u^T(T)) + O(e^{-\lambda_1 T/2}),$$

$$\frac{d}{d\tau}\bigg|_{\tau=T}(s^\tau(T-\tau), u^\tau(T-\tau)) = X_N(s^T(0), u^T(0)) + O(e^{-\lambda_1 T/2}),$$

$$\tag{10.3}$$

where X_N denotes the Hamiltonian vector field of N.

Proof. Consider two solutions (s^{T_1}, u^{T_1}) and (s^{T_2}, u^{T_2}) of the boundary value problem, we extend the definition of the solutions to $\mathbb{R}$ by solving the ODE. We note that $(s^{T_1}, u^{T_1}) : [0, T_2] \to \mathbb{R}^4$ satisfies the boundary condition s^{in} and $u^{T_1}(T_2)$ on $[0, T_2]$, and by Corollary 10.3,

$$\left\| (s^{T_1}, u^{T_1})(t) - (s^{T_2}, u^{T_2})(t) \right\| \leq C e^{\lambda_1(T_2-t)/2} |u^{T_1}(T_2) - u^{out}|.$$

Note that

$$\lim_{T_2 \to T_1} \frac{u^{T_1}(T_2) - u^{out}}{T_2 - T_1} = \lim_{T_2 \to T_1} \frac{u^{T_1}(T_2) - u^{T_1}(T_1)}{T_2 - T_1} = \frac{d}{dt}\bigg|_{t=T_1} u^{T_1}(t),$$

and as a result,

$$\left\| \frac{d}{dT}\Big|_{T=T_1} (s^T, u^T)(t) \right\| \leq Ce^{\lambda_1(T_1-t)}.$$

This proves the first half of (10.2) while the second half is similar.

For (10.3), note

$$\frac{d}{d\tau}\Big|_{\tau=T} (s^\tau(T), u^\tau(T)) = \frac{d}{d\tau}\Big|_{\tau=T} (s^\tau(\tau), u^\tau(\tau)) - \frac{d}{d\tau}\Big|_{\tau=T} (s^T(\tau), u^T(\tau)),$$

the first half of (10.3) follows. The second half is similar. □

Let $(v_{s_1}, v_{s_2}, v_{u_1}, v_{u_2})$ denote the coordinates for the tangent space induced by (s_1, s_2, u_1, u_2). For $K > 0$ and $x \in B_r$, we define the *strong unstable cone* by

$$C_K^u(x) = \{K^2|v_{u_2}|^2 > |v_{u_1}|^2 + |v_{s_1}|^2 + |v_{s_2}|^2\} \tag{10.4}$$

and the *strong stable cone* to be

$$C_K^s = \{K^2|v_{s_2}|^2 > |v_{s_1}|^2 + |v_{u_1}|^2 + |v_{u_2}|^2\}.$$

The following statement follows from the hyperbolicity of the fixed point O via standard techniques.

Lemma 10.4 (See for example [49]). *For any $0 < \kappa < \lambda_2 - \lambda_1$, there exists $\delta = \delta(\kappa, K)$ and $C = C(\kappa, K) > 1$ such that the following hold:*

- *If $\phi_N^t(x) \in B_{4\delta}(O)$ for $0 \leq t \leq t_0$ (ϕ_N^t is the Hamiltonian flow of the normal form Hamiltonian N), then $D\phi_N^t(C_K^u(x)) \subset C_K^u(\phi_N^t(x))$ for all $0 \leq t \leq t_0$. Furthermore, for any $v \in C_K^u(x)$,*

$$\|D\phi_N^t(x)v\| \geq C^{-1}e^{(\lambda_2-\kappa)t}, \quad 0 \leq t \leq t_0.$$

- *If $\phi_N^{-t}(x) \in B_{4\delta}(O)$ for $0 \leq t \leq t_0$, then $D\phi_N^{-t}(C_K^s(x)) \subset C_K^s(\phi_N^{-t}(x))$ for all $0 \leq t \leq t_0$. Furthermore, for any $v \in C_K^s(x)$,*

$$\|D\phi_N^{-t}(x)v\| \geq C^{-1}e^{(\lambda_2-\kappa)t}, \quad 0 \leq t \leq t_0.$$

Lemma 10.5. *For any $\kappa > 0$ there is $\delta_0 > 0$ such that for any $0 < \delta \leq \delta_0$, $\|s_1^{in}\| = \|u_1^{out}\| = \delta$, and $\|s_2^{in}\|, \|u_2^{out}\| \leq \kappa\lambda_1\delta/(2\lambda_2)$, we have*

$$\left\| \frac{d}{dT}u_2^T(0) \right\| \leq \kappa \left\| \frac{d}{dT}u_1^T(0) \right\|, \quad \left\| \frac{d}{dT}s_2^T(T) \right\| \leq \kappa \left\| \frac{d}{dT}s_1^T(T) \right\|.$$

Moreover, by integrating in T, we get

$$\|u_2^T(0)\| \leq \kappa\|u_1^T(0)\|, \quad \|s_2^T(T)\| \leq \kappa\|s_1^T(T)\|.$$

Proof. Given $\kappa > 0$ we can choose δ_0 small enough such that the backward flow

$D\phi_{-t}$ preserves the cone

$$(C_K^u)^c = \{|v_{u_2}| \leq K^{-1}\|(v_{s_1}, v_{s_2}, v_{u_1})\|\},$$

where $K = \kappa^{-1}$. Note that

$$\frac{d}{dT}(s^T(0), u^T(0)) = D\phi_{-T}\frac{d}{d\tau}\Big|_{\tau=T}(s^\tau(T), u^\tau(T)).$$

We have

$$\begin{aligned}
\frac{d}{d\tau}\Big|_{\tau=T}(s^\tau(T), u^\tau(T)) &= -X_N(s^T(T), u^T(T)) + O(e^{-\lambda_1 T/2}) \\
&= (\lambda_1 s_1^T(T), \lambda_2 s_2^T(T), \lambda_1 u_1^T(T), \lambda_2 u_2^T(T)) + O(\delta^2) + O(e^{-\lambda_1 T/2}) \\
&= (0, 0, \lambda_1\delta, \lambda_2 u_2^{out}) + O(\delta^2) + O(e^{-\lambda_1 T/2}) \in (C_K^u)^c,
\end{aligned}$$

if $|u_2^{out}| \leq \kappa\lambda_1\delta/(2\lambda_2)$, δ is small and T is large enough. By the invariance of unstable cones, $\frac{d}{dT}(s^T(0), u^T(0)) \in (C_K^u)^c$. Keep in mind that $\frac{d}{dT}s^T(0) = 0$, the first half of our estimate follows. The second half is proven in a symmetric way.   □

10.3   PROPERTIES OF THE LOCAL MAPS

Consider first the simple critical case, where we have a unique shortest loop γ_h^+ for the Jacobi matrix. The corresponding Hamiltonian orbit η_h^+ is a non-degenerate homoclinic orbit. We drop the subscript h and call η^- the time-reversal of η^+.

$$\begin{aligned}
\Sigma_+^s &= \{(\delta, s_2, u_1, u_1) : |s_2|, |u_1|, |u_2| \leq \delta\}, \\
\Sigma_+^u &= \{(s_1, s_2, \delta, u_1) : |s_2|, |u_1|, |u_2| \leq \delta\}.
\end{aligned} \tag{10.5}$$

Denote $q^+ = \eta^+ \cap \{s_1 = \delta\} = \{(\sigma, s_2^+, 0, 0)\}$, and $p^+ = \eta^+ \cap \{u_1 = \delta\} = \{(0, 0, \delta, u_2^+)\}$. Let $\kappa > 0$ be a parameter to be defined in Proposition 10.7. Since η^+ is tangent to the s_1, u_1 axes, we can choose $\delta > 0$ such that $|s_2^+|, |u_2^+| < \kappa\delta$, therefore q^+, p^+ are contained Σ_+^s and Σ_+^u respectively. Define

$$\begin{aligned}
l^s &= \{(\delta, s_2, 0, 0) : -\kappa\delta \leq s_2 \leq \kappa\delta\} \subset W^s(O) \cap \Sigma_+^s, \\
l^u &= \{(0, 0, \delta, u_2) : -\kappa\delta \leq u_2 \leq \kappa\delta\} \subset W^u(O) \cap \Sigma_+^u.
\end{aligned}$$

Let

$$\Sigma_+^{s,E} = \Sigma_+^s \cap \{N(s, u) = E\}, \quad \Sigma_+^{u,E} = \Sigma_+^u \cap \{N(s, u) = E\}$$

be the restriction of the sections to an energy E close to 0. We would like to

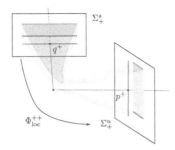

Figure 10.1: Rectangles mapped under Φ_{loc}^{++}

study the domain of the restricted local map $\Phi_{loc}^{++} : \Sigma_+^{s,E} \to \Sigma_+^{u,E}$. The following lemma allows us to parameterize both sections using the s_2, u_2 variables.

Lemma 10.6. *There exists $\kappa \in (0,1)$ such that for $|E| < \kappa\delta$, the set $\Sigma_+^{s,E} \cap \{|s_2|, |u_2| < \kappa\delta\}$ is given by a graph $u_1 = u_1^{E,s}(s_2, u_2)$ with $\|Du_1^{E,s}\| = O(1)$. The same holds for $\Sigma_+^{u,E}$ with the graph given by $s_1 = s_1^{E,u}$.*

Proof. We restrict to $(s, u) \in \Sigma_+^s$ and set $s_1 = \delta$. Since $N(s, u) = \lambda_1 \delta u_1 + \lambda_2 s_2 u_2 + O(\delta)$, for sufficiently small κ, $N(s_1, \delta, s_2, u_2) - E > 0$, $N(s_1, -\delta, s_2, u_2) - E < 0$, $\partial_{u_1} N > 0$, implying the existence of a unique u_1 solving $N - E = 0$, which we denote $u_1^{E,s}(s_2, u_2)$. As a function, $u_1^{E,s}$ is automatically C^{r-2} (since N is C^{r-1}) by the implicit function theorem, and the norm estimates for its derivative follows a direct computation. The case for $\Sigma_+^{u,E}$ is similar. $\square$

A rectangle R is a diffeomorphic image of the Euclidean rectangle in $\mathbb{R}^2$. Let us label the vertices by $1, 2, 3, 4$ in clockwise order, and call the four sides $l_{12}, l_{34}, l_{14}, l_{23}$.

Proposition 10.7. *There exist $e, \kappa > 0$ such that for each $0 < E < e$, there is a rectangle $R^{++}(E) \subset \Sigma_+^{s,E}$ with sides $l_{ij}(E)$, such that:*

1. *Φ_{loc}^{++} is well defined on $R^{++}(E)$, and its image $\Phi_{loc}^{++}(R^{++}(E))$ is also a rectangle, it's four sides denoted $l'_{ij}(E)$.*
2. *As $E \to 0$, both $l_{12}(E)$ and $l_{34}(E)$ converge to l^u in Hausdorff metric. Similarly, both $l'_{14}(E)$ and $l'_{23}(E)$ converge to l^u.*
3. *If $(s, u) \in \Sigma_+^{s,E}$, $(s', u') \in \Sigma_+^{u,E}$ is such that $\Phi_{loc}^{++}(s, u) = (s', u')$, with $(s, u) \in B_{\kappa\delta}(q^+)$ and $(s', u') \in B_{\kappa\delta}(p^+)$, then $(s, u) \in R^{++}(E)$.*

See Figure 10.1.

We first prove a version of Proposition 10.2 using E as a parameter.

Proposition 10.8. *There is $e, \kappa > 0$ such that if s^{in} and u^{out} satisfies*

$$|s_2^{in}|, |u_2^{out}| \le \frac{\lambda_1}{2\lambda_2}\kappa\delta, \quad s_1^{in} = u_1^{out} = \delta,$$

then for all $E \in (0, e)$ there is $T = T_E > 0$ and a unique orbit (s^E, u^E) : $[0, T_E] \to B_{4\delta}(O)$, such that

$$s^E(0) = s^{in}, \quad u^E(T_E) = u^{out}.$$

Proof of Proposition 10.8. Let S_E denote the energy surface $\{N(s, u) = E\}$. Given s^{in} and s^{out} and $T > T_0$, Proposition 10.2 implies the existence of a solution (s^T, u^T) solving the boundary value problem. Moreover, as $T \to \infty$, $(s^T, u^T)(0) \to (s^{in}, 0) \in S_0$. Writing $E(T)$ the energy of the orbit (s^T, u^T), we have $E(T) \to 0$ as $T \to \infty$. The conclusions of our proposition follow from Proposition 10.2, as long as we can show the following claims.

Claim: (1) $E(T)$ is positive and strictly monotone; therefore $E(T)$ is one-to-one and onto for $T \in [T_0, \infty)$; (2) There is $e > 0$ such that uniformly over all s^{in}, u^{out}, we have $0 < E(T) < e$ if $T \ge T_0$, and therefore the inverse T_E is well defined for $E \in (0, e]$.

By Lemma 10.5 $|u_2^T(0)| \le \kappa |u_1^T(0)|$. We first show $N(s^T(0), u^T(0)) > 0$. Setting $E = 0$, Lemma 10.6 implies there exists a graph $u_1 = u_1^{0,s}(s_2, u_2)$ on which $N = 0$. This submanifold divides $\Sigma^+ \cap \{|s_2|, |u_2| < \kappa\delta\}$ into two components, with

- $N(s, u) > 0$ whenever $u_1 > u_1^{0,u}(s_2, u_2)$, and $N(s, u) < 0$ whenever $u_1 < u_1^{0,u}(s_2, u_2)$.

Moreover, $u_1^{0,u}|_{u_2=0} = 0$, and $|\partial_{u_2} u_1^c| \le C$, where C is a constant depending on $\|N\|_{C^2}$. Therefore $|u_1^c(s_2, u_2)| \le C|u_2|$. This implies if $(\delta, u_1, s_2, u_2) \in \Sigma$ satisfies $u_1 > C|u_2| \ge |u_1^{0,u}|$, we have $N(\delta, u_1, s_2, u_2) > 0$. In particular, this is satisfied for $(s^T(0), u^T(0))$ if $\kappa < C^{-1}$.

$E(T)$ is a continuous function defined on $T \ge T_0$, and $\lim_{T\to\infty} E(T) = 0$. We now prove that it is monotone. Using Lemma 10.5,

$$\frac{d}{dT}E(T) = \frac{d}{dT}N(s^T(0), u^T(0)) = \nabla N \cdot \left(0, 0, \frac{d}{dT}u_1^T(0), \frac{d}{dT}u_2^T(0)\right)$$

$$= (\lambda_1 s_1^{in} + O_2(s, u))\frac{d}{dT}u_1^T(0) + (\lambda_2 s_2^{in} + O_2(s, u))\frac{d}{dT}u_2^T(0) \ne 0$$

if κ is sufficiently small. Since we have proved $E(T) > 0$, claim (1) follows. Finally, Proposition 10.2 implies

$$0 < E(T) \le C\delta^2 e^{-\lambda_1 T/2},$$

and claim (2) follows. □

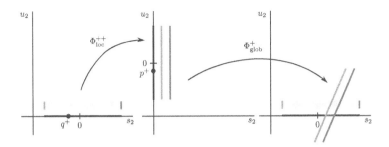

Figure 10.2: Local map and global map in s_2, u_2 coordinates.

Proof of Proposition 10.7. Let $\kappa > 0$ be small enough such that Proposition 10.8 applies, and let us rename $\lambda_1 \kappa / (2\lambda_2)$ into κ. Consider two parameters $a \in [-\kappa\delta, \kappa\delta]$ and $b \in [-\kappa\delta, \kappa\delta]$, and then there exist unique orbits $s_{a,b}^E$ solving the boundary value problem

$$s_{a,b}^E(0) = (\delta, a), \quad s_{a,b}^E(T_{a,b}(E)) = (\delta, b),$$

then

$$R^{++}(E) = \left\{ (s_{a,b}^E(0), u_{a,b}^E(0)) : a, b \in [\delta/2, \delta/2] \right\},$$
$$\Phi_{\text{loc}}^{++} \left(R^{++}(E) \right) = \left\{ (s_{a,b}^E(T_{a,b}(E)), u_{a,b}^E(T_{a,b}(E))) : a, b \in [\delta/2, \delta/2] \right\}.$$

An illustration of the rectangles in the (s_2, u_2) variables is in Figure 10.2.

Note that two sides of the rectangle $R^{++}(E)$ are graphs over $s_2 \in [-\kappa\delta, \kappa\delta]$; since the u_1, u_2 components converge to 0 exponentially fast as $T \to \infty$ (and $E \to 0$), we conclude that these two sides converge to l^s. The same can be said about the rectangle $\Phi_{\text{loc}}^{++} \left(R^{++}(E) \right)$ and l^u.

Finally, note that by definition, $R^{++}(E)$ contains the initial point of all orbits $(s, u) : [0, T] \to S_E$ such that $s_1(0) = \delta$, $|s_2(0)| \le \kappa\delta$, $u_1(T) = \delta$, and $|u_2(T)| \le \kappa\delta$, which contains the orbits such that $(s, u)(0) \in B_{\kappa\delta}(p)$ and $(s, u)(T) \in B_{\kappa\delta}(p^+)$. ☐

For each of the other symbols $--$, $+-$, and $-+$, statements analogous to Proposition 10.7 hold, after making appropriate changes. One important point is that we consider $E \in (0, e)$ (positive energy) for the symbols $++$ and $--$, and $E \in (-e, 0)$ (negative energy) for the symbols $+-$ and $-+$.

We now turn to the global map. We have $p^+ \in l^u$, $q^+ \in l^s$, and $\Phi_{\text{glob}}^+(p^+) = q^+$. Moreover, condition $[DR4^c]$ implies that $\Phi_{\text{glob}}^+(l^u)$ intersects l^s transversally in the energy surface S_E. Since the transition time for the Poincaré map Φ_{glob}^+ is uniformly bounded, the restricted map $\Phi_{\text{glob}}^+ | S_E$ depends smoothly on E. By Proposition 10.7, for sufficiently small E, the rectangle $\Phi_{\text{glob}}^+ \circ \Phi_{\text{loc}}^{++}(R^{++}(E))$ intersects $R^{++}(E)$ transversally. Applying the same argument to the other symbols, we obtain the following corollary.

Corollary 10.9. *There exists $e > 0$ such that the following hold.*

1. *For $0 < E < e$, $\Phi^+_{\text{glob}} \circ \Phi^{++}_{\text{loc}}(R^{++}(E))$ intersects $R^{++}(E)$ transversally. This means the images of l_{12} and l_{34} intersect l_{12} and l_{34} transversally, and the images of γ_{14} and γ_{23} do not intersect $R^{++}(E)$.*
2. *For $0 < E < e$, $\Phi^-_{\text{glob}} \circ \Phi^{--}_{\text{loc}}(R^{--}(E))$ intersects $R^{--}(E)$ transversally.*
3. *For $-e < E < 0$, $\Phi^-_{\text{glob}} \circ \Phi^{+-}_{\text{loc}}(R^{+-}(E))$ intersect $R^{-+}(E)$ transversally, and $\Phi^+_{\text{glob}} \circ \Phi^{-+}_{\text{loc}}(R^{-+}(E))$ intersect $R^{+-}(E)$ transversally.*

See Figure 10.2 for a demonstration.

10.4   PERIODIC ORBITS FOR THE LOCAL AND GLOBAL MAPS

The rectangles as constructed form isolating blocks of Conley and McGehee [68]. We call a rectangle $R = I_1 \times I_2 \subset \mathbb{R}^d \times \mathbb{R}^k$, $I_1 = \{\|x_1\| \leq 1\}$, $I_2 = \{\|x_2\| \leq 1\}$ *an isolating block* for the C^1 diffeomorphism Φ, if the following hold:

1. The projection of $\Phi(R)$ to the first component covers I_1.
2. $\Phi|I_1 \times \partial I_2$ is homotopically equivalent to the identity restricted on the set $I_1 \times (\mathbb{R}^k \setminus \text{int } I_2)$.

See Figure 10.3.

Proposition 10.10. *Let R be an isolating block for the map Φ. Then the sets*

$$W^+ = \{x \in R : \Phi^k(x) \in R, \ k \geq 0\}, \quad W^- = \{x \in R : \Phi^{-k}(x) \in R, \ k \geq 0\})$$

are non-empty, W^+ project onto I_1, W^- project onto I_2, and $W^+ \cap W^-$ is non-empty.

We now apply the isolating block construction to the maps and rectangles obtained in Corollary 10.9.

Proposition 10.11. *There exists $e > 0$ such that the following hold.*

1. *For $0 < E < e$, $\Phi^+_{\text{glob}} \circ \Phi^{++}_{\text{loc}}$ has a unique fixed point $p^+(E)$ on $\Sigma^s_+ \cap R^{++}(E)$.*
2. *For $0 < E < e$, $\Phi^-_{\text{glob}} \circ \Phi^{--}_{\text{loc}}$ has a unique fixed point $p^-(E)$ on $\Sigma^s_- \cap R^{--}(E)$.*
3. *For $-e < E < 0$: $\Phi^+_{\text{glob}} \circ \Phi^{-+}_{\text{loc}} \circ \Phi^-_{\text{glob}} \circ \Phi^{+-}_{\text{loc}}$ has a unique fixed point $p^c(E)$ on $R^{+-}(E) \cap (\Phi^-_{\text{glob}} \circ \Phi^{+-}_{\text{loc}})^{-1}(R^{-+}(E))$.*

Notice that the rectangle $R^{++}(T)$ has C^1 sides, and there exists a C^1 change of coordinates turning it to a standard rectangle. It's easy to see that the

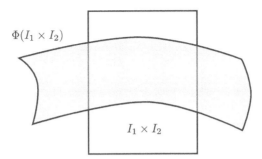

Figure 10.3: Isolating block.

isolating block conditions are satisfied for the following maps and rectangles:

$$\Phi^+_{\text{glob}} \circ \Phi^{++}_{\text{loc}} \quad \text{and} \quad R^{++}(E), \qquad \Phi^-_{\text{glob}} \circ \Phi^{--}_{\text{loc}} \quad \text{and} \quad R^{--}(E),$$

$$\Phi^+_{\text{glob}} \circ \Phi^{-+}_{\text{loc}} \circ \Phi^-_{\text{glob}} \circ \Phi^{+-}_{\text{loc}} \quad \text{and} \quad (\Phi^-_{\text{glob}} \circ \Phi^{+-}_{\text{loc}})^{-1} R^{-+}(E) \cap R^{+-}(E).$$

We will only prove item 1 of Proposition 10.11, as the rest are analogous. By Proposition 10.10, we obtain that

$$W^+(E) = \bigcap_{k \geq 0} (\Phi^+_{\text{glob}} \circ \Phi^{++}_{\text{loc}})^k R^{++}(E), \quad W^-(E) = \bigcap_{k \geq 0} (\Phi^+_{\text{glob}} \circ \Phi^{++}_{\text{loc}})^{-k} R^{++}(E)$$

are non-empty. In order to prove that the intersection $W^+(E) \cap W^-(E)$ is a single point, we use the invariant cones introduced in (10.4).

Lemma 10.12. *There exists $K > 1$ and $T_0 > 0$ such that the following hold. Assume that $U \subset \Sigma^s_+ \cap B_r$ is a connected open set on which the local map Φ^{++}_{loc} is defined, and for each $x \in U$,*

$$\inf\{t \geq 0 : \phi^t_N(x) \in \Sigma^u_+\} \geq T_0.$$

Then the map $D(\Phi^+_{\text{glob}} \circ \Phi^{++}_{\text{loc}})$ preserves the cone field C^u_K, and the inverse $D(\Phi^+_{\text{glob}} \circ \Phi^{++}_{\text{loc}})^{-1}$ preserves C^s_K.

Recall that $l^u \ni p^+$ is the intersection of the unstable manifold $W^u(O)$ with the section Σ^u_+. Let T^u be the tangent vector to l^u at p^+, and T^s the tangent vector to T^s at q^+. We will show that if $\Phi^{++}_{\text{loc}}|_{S_E}(x) = y$, then the image of the unstable cone $D\Phi^{++}_{\text{loc}}(x)C^u_K$ is very close to T^u. This happens because the flow of tangent vector is very close to that of a linear flow.

The following lemma holds for a general flow ϕ^t on $\mathbb{R}^d \times \mathbb{R}^k$, and a trajectory x_t of the flow. Let $v(t) = (v_1(t), v_2(t))$ be a solution of the variational equation, i.e. $v(t) = D\phi^t(x_t)v(0)$.

Lemma 10.13. *With the above notations assume that there exist $b_2 > 0$, $b_1 <$*

b_2, and $\sigma, \delta > 0$ such that the variational equation

$$\dot{v}(t) = \begin{bmatrix} A(t) & B(t) \\ C(t) & D(t) \end{bmatrix} \begin{bmatrix} v_1(t) \\ v_2(t) \end{bmatrix}$$

satisfies $A \leq b_1 I$ and $D \geq b_2 I$ as quadratic forms, and $\|B\| \leq \sigma$, $\|C\| \leq \delta$.

Then for any $c > 0$ and $\kappa > 0$, there exists $\delta_0 > 0$ such that if $0 < \delta, \sigma < \delta_0$, we have

$$(D\phi^t) C_K^u \subset C_{1/\beta_t}^u, \quad \beta_t = ce^{-(b_2 - b_1 - \kappa)t} + \sigma/(b_2 - b_1 - \kappa).$$

Proof. Denote $\gamma_0 = c$. The invariance of the cone field is equivalent to

$$\frac{d}{dt} \left(\beta_t^2 \langle v_2(t), v_2(t) \rangle - \langle v_1(t), v_1(t) \rangle \right) \geq 0.$$

Computing the derivatives using the variational equation, and applying the norm bounds and the cone condition, we obtain

$$2\beta_t \left(\beta_t' + (b_2 - \delta\beta_t - b_1)\beta_t - \sigma \right) \|v_2\|^2 \geq 0.$$

We assume that $\beta_t \leq 2\gamma_0$, then for sufficiently small δ_0, $\delta\beta_t \leq \kappa$. Denote $b_3 = b_2 - b_1 - \kappa$ and let β_t solve the differential equation

$$\beta_t' = -b_3\beta_t + \sigma.$$

It's clear that the inequality is satisfied for our choice of β_t. Solve the differential equation for β_t and the lemma follows. $\qquad\square$

Proof of Lemma 10.12. We will only prove the unstable version. By Assumption 4, there exists $c > 0$ such that $D\Phi_{\text{glob}}^+(q^+)T^{uu}(q^+) \subset C_K^u(p^+)$. Note that as $T_0 \to \infty$, the neighborhood U shrinks to p^+ and V shrinks to q^+. Hence there exist $\beta > 0$ and $T_0 > 0$ such that $D\Phi_{\text{glob}}^+(y)C_{1/\beta}^u(y) \subset C_K^u$ for all $y \in V$.

Let $(s, u)(t)_{0 \leq t \leq T}$ be the trajectory from x to y. By Proposition 10.2, we have $\|s\| \leq e^{-\lambda_1 T/4}$ for all $T/2 \leq t \leq T$. It follows that the matrix for the variational equation

$$\begin{bmatrix} A(t) & B(t) \\ C(t) & D(t) \end{bmatrix} = \begin{bmatrix} -\operatorname{diag}\{\lambda_1, \lambda_2\} + O(s) & O(s) \\ O(u) & \operatorname{diag}\{\lambda_1, \lambda_2\} + O(u) \end{bmatrix} \quad (10.6)$$

satisfies $A \leq -(\lambda_1 - \kappa)I$, $D \geq (\lambda_1 - \kappa)I$, $\|C\| = O(\delta)$, and $\|B\| = O(e^{-(\lambda_1 - \kappa)T/2})$. As before, $C_K^u(x) = \{\|v_s\| \leq c\|v_u\|\}$, and Lemma 10.13 implies

$$D\phi_T(x)C_K^u(x) \subset C_{1/\beta_T}^u(y),$$

where $\beta_T = O(e^{-\lambda' T/2})$ and $\lambda' = \min\{\lambda_2 - \lambda_1 - \kappa, \lambda_1 - \kappa\}$. Finally, note that $D\phi_T(x)C_K^u(x)$ and $D\Phi_{\text{loc}}^{++}(x)C_K^u(x)$ differ by the differential of the local

Poincaré map near y. Since near y we have $|s| = O(e^{-(\lambda_1 - \kappa)T})$, using the equation of motion, the Poincaré map is exponentially close to identity on the (s_1, s_2) components, and is exponentially close to a projection of u_2 on the (u_1, u_2) components. It follows that the cone C^u_{1/β_T} is mapped by the Poincaré map into a strong unstable cone with exponentially small size. In particular, for $T \geq T_0$, we have

$$D\Phi^{++}_{\text{loc}}(x) C^u_K(x) \subset C^u_{1/\beta}(y),$$

and the lemma follows. ☐

Proof of Proposition 10.11. We have proven that the map $D(\Phi^+_{\text{glob}} \circ \Phi^{++}_{\text{loc}})$ preserves the unstable cone field C^u_K. Similar to the argument in Lemma 10.5, this implies any two points $z_1, z_2 \in W^+(E)$ must satisfy $z_1 \in z_2 + (C^u_K)^c$, otherwise, we can use the expansion in the u_2 direction to show that the distance between the forward orbits of z_1 and z_2 must grow exponentially, contradicting the fact that $z_1, z_2 \in W^+(E)$. Moreover, in Lemma 10.6 we showed that $\Sigma^{s,E}_+$ can be parameterized by s_2, u_2, this implies $W^+(E)$ is a K^{-1}-Lipschitz graph over s_2. By the same argument, $W^-(E)$ is a K^{-1}-Lipschitz graph over u_2. Since $K > 1$, there is a unique intersection in $W^+(E) \cap W^-(E)$, which we denote by $p^+(E)$. This concludes the proof of item 1. The other cases are analogous. ☐

In the case of the non-simple homology, there exist two rectangles R_1 and R_2, whose images under $\Phi_{\text{glob}} \circ \Phi_{\text{loc}}$ intersect themselves transversally, providing a "horseshoe" type picture.

Proposition 10.14. *There exists $e > 0$ such that the following hold:*

1. *For all $0 < E \leq e$, there exist rectangles $R_1(E), R_2(E) \in \Sigma^{s,E}_+$ such that for $i = 1, 2$, $\Phi_{\text{glob}} \circ \Phi^{++}_{\text{loc}}(R_i)$ intersects both $R_1(E)$ and $R_2(E)$ transversally.*
2. *Given $\sigma = (\sigma_1, \cdots, \sigma_n)$, there exists a unique fixed point $p^\sigma(E)$ of*

$$\prod_{i=n}^{1} \left(\Phi^{\sigma_i}_{\text{glob}} \circ \Phi^{++}_{\text{loc}} \right) |_{R_{\sigma_i}(E)} \tag{10.7}$$

   on the set $R_{\sigma_1}(E)$.
3. *The curve $p^\sigma(E)$ is a C^1 graph over the u_1 component with uniformly bounded derivatives. Furthermore, $p^\sigma(E)$ approaches p^{σ_1} and for each $1 \leq j \leq n - 1$,*

$$\prod_{i=j}^{1} \left(\Phi^{\sigma_i}_{\text{glob}} \circ \Phi^{++}_{\text{loc}} \right) (p^\sigma(E))$$

   approaches $p^{\sigma_{j+1}}$ as $E \to 0$.

The proof is analogous to that of Proposition 10.11.

10.5   NORMALLY HYPERBOLIC INVARIANT MANIFOLDS

We prove Theorem 4.4 in this section.

Proof. Non-simple case. If the homology h is non-simple, then $h = n_1 h_1 + n_2 h_2$ with h_1, h_2 being simple homologies. Let $(\sigma_1, \ldots, \sigma_n)$ be the sequence determined by Lemma 4.2. Applying Proposition 10.14, we obtain the fixed points $p^\sigma(E)$ for all $0 < E < e$. The fixed points correspond to hyperbolic periodic orbits that we call η_h^E. Let γ_h^E be the projection of η_h^E to the configuration space, we now prove that they must be identical to the shortest curve in Jacobi metric g_E, after a reparametrization. According to the condition $[DR3^c]$, γ_h^0 is the unique shortest curve for the Jacobi metric g_0, and there exists c_0 such that $\eta_h^0 = \widetilde{\mathcal{A}}_{H^s}(c_0) = \widetilde{\mathcal{N}}_{H^s}(c_0)$. Any g_E shortest curve γ_E' corresponds to the Aubry set of cohomology c_E, which lifts to an orbit η_E' in phase space. Using semi-continuity, η_E' must be contained in a neighborhood of η_h^0 in the phase space. In particular, it must intersect the sections Σ_+^s and Σ_+^u sufficiently close to q^+ and p^+. According to Proposition 10.8, item 3, the intersection with Σ_+^s must be contained in the rectangle $R^{++}(E)$. Since the fixed point $p^\sigma(E)$ is the unique fixed point for the map (10.7), we conclude that $\eta_E' = \eta_h^E$.

Simple critical case. The existence of the periodic orbits follows from Proposition 10.11 in the same way as the non-simple case. Also, by the same reasoning, we know that the orbits η_h^E, η_{-h}^E must coincide with the minimal geodesics of the Jacobi metric. It suffices to show that

$$\mathcal{M} = \bigcup_{0 < E < e} (\eta_h^E \cup \eta_{-h}^E) \cup \eta_h^0 \cup \eta_{-h}^0 \cup \bigcup_{-e < E < 0} \eta_c^E$$

form a C^1 normally hyperbolic invariant cylinder.

Denote

$$l^+(p^+) = \{p^+(E)\}_{0 < E \leq E_0}, \quad l^+(p^-) = \{p^-(E)\}_{0 < E \leq E_0},$$

$l^+(q^+) = \Phi_{\text{loc}}^{++}(l^+(p^+))$ and $l^+(q^-) = \Phi_{\text{loc}}^{--}(l^+(q^-))$. Note that the superscript of l indicates positive energy instead of the signature of the homoclinics. We denote

$$l^-(p^+) = \{p^c(E)\}_{-E_0 \leq E < 0}$$

$l^-(q^-) = \Phi_{\text{loc}}^{+-}(l^-(p^+))$, $l^-(p^-) = \Phi_{\text{glob}}^{--}(l^-(q^-))$ and $l^-(q^+) = \Phi_{\text{loc}}^{-+}(l^-(p^-))$. An illustration of $\mathcal{M}$ and the curves $l^\pm$ are included in Figure 10.4.

We now define $l(p^\pm) = l^-(p^\pm) \cup \{p^\pm\} \cup l^-(p^\pm)$. By considering the image $\bigcup_{t>0} \phi_t(l(p^+))$ and $\bigcup_{t>0} \phi_t(l(q^+))$, the projection of $\mathcal{M}$ to the $s_1 u_1$ plane contains a neighborhood of 0. Invariance of the unstable cones C_K^u implies that $\mathcal{M}$ is a Lipschitz graph over $s_1 u_1$ near 0. It follows that $\mathcal{M}$ is a *center manifold*, which is $C^{1+\alpha}$ for α depending on the spectral gap between the stable/unstable direction and the central direction (see [49]).

Finally, note that all the arguments apply to the normal form system (10.1).

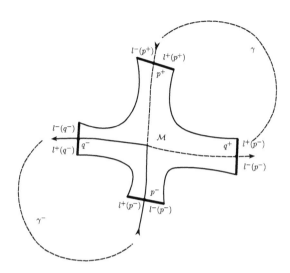

Figure 10.4: Invariant manifold $\mathcal{M}$ near the origin.

The same conclusions hold for small C^2 perturbation to the Hamiltonian (which leads to a small C^1 perturbation of the normal form). □

10.6   CYCLIC CONCATENATIONS OF SIMPLE GEODESICS

We provide the proof of the auxiliary result Lemma 4.2 before proceeding to the next chapter.

Denote $\gamma_1 = \gamma_0^{h_1}$ and $\gamma_2 = \gamma_0^{h_2}$ and $\gamma = \gamma_h^0$. Recall that γ has homology class $n_1 h_1 + n_2 h_2$ and is the concatenation of n_1 copies of γ_1 and n_2 copies of γ_2. Since h_1 and h_2 generate $H_1(\mathbb{T}^2, \mathbb{Z})$, by introducing a linear change of coordinates, we may assume $h_1 = (1, 0)$ and $h_2 = (0, 1)$.

Given $y \in \mathbb{T}^2 \setminus \gamma \cup \gamma_1 \cup \gamma_2$, the fundamental group of $\mathbb{T}^2 \setminus \{y\}$ is a free group of two generators, and in particular, we can choose γ_1 and γ_2 as generators. (We use the same notations for the closed curves γ_i, $i = 1, 2$ and their homotopy classes.) curve γ determines an element

$$\gamma = \prod_{i=1}^{n} \gamma_{\sigma_i}^{s_i}, \quad \sigma_i \in \{1, 2\}, \ s_i \in \{0, 1\}$$

of this group. Moreover, the translation $\gamma_t(\cdot) := \gamma(\cdot + t)$ of γ determines a new element by cyclic translation, i.e.,

$$\gamma_t = \prod_{i=1}^{n} \gamma_{\sigma_{i+m}}^{s_{i+m}}, \quad m \in \mathbb{Z},$$

where the sequences σ_i and s_i are extended periodically. We claim the following:

There exists a unique (up to translation) periodic sequence σ_i such that $\gamma = \prod_{i=1}^{n} \gamma_{\sigma_{i+m}}$ for some $m \in \mathbb{Z}$, independent of the choice of y. Note that in particular, all $s_i = 1$.

The proof of this claim is split into two steps.

Step 1. Let $\gamma_{n_1/n_2}(t) = \{\gamma(0) + (n_1/n_2, 1)t, t \in \mathbb{R}\}$. We will show that γ is homotopic (along non-self-intersecting curves) to γ_{n_1/n_2}. To see this, we lift both curves to the universal cover with the notations $\tilde{\gamma}$ and $\tilde{\gamma}_{n_1/n_2}$. Let $p, q \in \mathbb{Z}$ be such that $pn_1 - qn_2 = 1$ and define

$$T\tilde{\gamma}(t) = \tilde{\gamma}(t) + (p, q).$$

As T generates all integer translations of $\tilde{\gamma}$, γ is non-self-intersecting if and only if $T\tilde{\gamma} \cap \tilde{\gamma} = \emptyset$. Defining the homotopy $\tilde{\gamma}_\lambda = \lambda\tilde{\gamma} + (1 - \lambda)\tilde{\gamma}_{n_1/n_2}$, it suffices to prove $T\tilde{\gamma}_\lambda \cap \tilde{\gamma}_\lambda = \emptyset$. Take an additional coordinate change

$$\begin{bmatrix} x \\ y \end{bmatrix} \mapsto \begin{bmatrix} n_1 & p \\ n_2 & q \end{bmatrix}^{-1} \begin{bmatrix} x \\ y \end{bmatrix};$$

then under the new coordinates $T\tilde{\gamma}(t) = \tilde{\gamma}(t) + (1, 0)$.

Under the new coordinates, $T\tilde{\gamma} \cap \tilde{\gamma} = \emptyset$ if and only if any two points on the same horizontal line have distance less than 1. The same property carries over to $\tilde{\gamma}_\lambda$ for $0 \le \lambda < 1$; hence $T\tilde{\gamma}_\lambda \cap \tilde{\gamma}_\lambda = \emptyset$.

Step 2. By step 1, it suffices to prove that $\gamma = \gamma_{n_1/n_2}$ defines unique sequences σ_i and s_i. Since $\tilde{\gamma}_{n_1/n_2}$ is increasing in both coordinates, we have $s_i = 1$ for all i. Moreover, choosing a different y is equivalent to shifting the generators γ_1 and γ_2. Since the translation of the generators is homotopic to identity, the homotopy class is not affected. This concludes the proof of Lemma 4.2.

Chapter Eleven

Aubry-Mather type at the double resonance

We consider the system

$$H^s(\varphi, I) = K(I) - U(\varphi), \quad \varphi \in \mathbb{T}^2, \quad I \in \mathbb{R}^2.$$

We assume that $\min U = 0$ and that the conditions $[\mathrm{DR}1^h] - [\mathrm{DR}3^h]$, $[\mathrm{DR}1^c]$ $- [\mathrm{DR}4^c]$ are satisfied. Given the homology $h \in H_1(\mathbb{T}^2, \mathbb{Z})$, let $\bar{c}_h : (0, \bar{E}] \to H^1(\mathbb{T}^2, \mathbb{R})$ be the curve of cohomology selected by Proposition 5.1. There are two regimes, the "high-energy regime," where we consider the cohomologies $\bar{c}_h(E)$ with $e \leq E \leq \bar{E}$ where e is a small parameter; and the critical regime, where c is in a small neighborhood of $\bar{c}_h(0)$.

11.1 HIGH-ENERGY CASE

First we consider the "non-critical energy case" and show that the cohomologies as chosen are of Aubry-Mather type. For each $E > 0$, there are two possible behaviors:

1. The Aubry set $\mathcal{A}_{H^s}(\bar{c}_h(E)) = \gamma_h^E$, where γ_h^E is the unique shortest geodesic in homology E. Let us denote the corresponding Hamiltonian orbit $\eta_h^E = (\varphi, I) : [0, T_E] \to \mathbb{T}^2 \times \mathbb{R}^2$.
2. (Bifurcation) The Aubry set $\mathcal{A}_{H^s}(\bar{c}_h(E)) = \gamma_h^E \cup \bar{\gamma}_h^E$, where $\gamma_h^E, \bar{\gamma}_h^E$ are the two shortest geodesics in homology E. Let us denote the corresponding Hamiltonian orbit $\eta_h^E, \bar{\eta}_h^E$.

Theorem 11.1. *Given any $e > 0$, there are $\epsilon_0, \delta > 0$ depending only on H^s and e such that the following holds for all $0 < \epsilon < \epsilon_0$ and all $U' \in V_\delta(U)$ (in the space $C^r(\mathbb{T}^2)$), and the Hamiltonian*

$$H_\epsilon^s(\varphi, I, \tau) = K(I) - U'(\varphi) + \sqrt{\epsilon}P, \quad \|P\|_{C^2} \leq 1,$$

satisfies the following properties:

1. *Suppose $E \geq e$ is such that $\mathcal{A}_{H^s}(\bar{c}_h(E))$ is a unique hyperbolic orbit, then $(H_\epsilon^s, \bar{c}_H(E))$ is of Aubry-Mather type.*
2. *Suppose $E \geq e$ is such that $\mathcal{A}_{H^s}(\bar{c}_h(E))$ is the union of two hyperbolic orbits; then $(H_\epsilon^s, \bar{c}_h(E))$ is of bifurcation Aubry-Mather type.*

We first discuss the non-bifurcation case. Given $E_0 \geq 0$, denote $c_0 = \bar{c}_h(E_0)$. Suppose $\mathcal{A}_{H^s}(c_0) = \gamma_h^{E_0}$ consists of a unique shortest curve; then there exists $E_1 < E_0 < E_2$ such that each η_h^E for $E \in (E_1, E_2)$ is a hyperbolic periodic orbit. Then

$$\mathcal{C}_0 = \bigcup_{E \in (E_1, E_2)} \eta_h^E \subset \mathbb{T}^2 \times \mathbb{R}^2$$

is a normally hyperbolic invariant cylinder. In order to give a proper parametrization for $\mathcal{C}_0$, we consider a transversal section Σ to $\eta_h^{E_0}$, which is also transversal to η_h^E, $E \in (E_1, E_2)$ if E_1, E_2 are close enough to E_0. Denote $z^E = \eta_h^E \cap \Sigma$, then Z^E, $E \in (E_1, E_2)$ is a smooth function of E. We now define

$$\chi : \mathbb{T} \times (E_1, E_2) \to \mathbb{T}^2 \times \mathbb{R}^2, \quad \chi(s, E) = \phi_{H^s}^{sT_E}(z^E), \tag{11.1}$$

where $\phi_{H^s}^t$ is the Hamiltonian flow of H^s and T_E is the period of η_h^E.

Before moving on, we recall the Green bundle introduced in Section 6.7. The main conclusion is that along a minimizing orbit $(\theta, p)(t)$, $t \in \mathbb{R}$, there exist two invariant bundles $\mathcal{G}_\pm$ given by Lipschitz graphs over the θ components. Moreover, if $(\theta, p)(t)$ is a hyperbolic orbit, then $\mathcal{G}_-$ coincide with the span of the unstable bundle and the flow direction, and $\mathcal{G}_+$ coincide with the span of the stable bundle and the flow direction (Proposition 6.19).

Lemma 11.2. *There is $\delta > 0$ depending on K, U, E_1, E_2 such that for all $U' \in \mathcal{V}_\delta(U)$ and $H^s = K - U'$, χ is a smooth embedding and $\chi(\mathbb{T} \times (E_1, E_2)) = \mathcal{C}_0$. $\mathcal{C}_0$ is normally hyperbolic, and for each E, $W^u(\eta_h^E)$ is a smooth graph over the θ component on a neighborhood of γ_h^E.*

Proof. The fact that γ_h^E is a unique non-degenerate shortest geodesic is robust, therefore χ is well defined for all $U' \in \mathcal{V}_\delta(U)$, where δ depends on K, U, E_1, E_2.

The fact that χ is smooth and that the image is $\mathcal{C}_0$ follows directly from the definition. Since $H^s(\chi(s, E)) = E$ by definition, we have $\nabla H^s(\chi(s, E)) \cdot \partial_E \chi(s, E) = 1$. On the other hand, $\partial_s \chi(s, E) = X_{H^s}(\chi(s, E))$, where X_{H^s} is the Hamiltonian vector field of H^s. Since $\nabla H^s \cdot X_{H^s} = 0$, we conclude that $\partial_s \chi, \partial_E \chi$ are linearly independent over $(s, E) \in \mathbb{T} \times (E_1, E_2)$.

To see that the local stable/unstable manifold are graphs over the θ component, we note that by Proposition 6.19, the Green bundles coincide with the stable/unstable bundle, and are Lipschitz graphs over the θ components. The Lipschitz constant is uniform over all $E > 0$. As a result, the unstable bundle of η_h^E projects onto a bundle transversal to γ_h^E, and the unstable manifold projects onto the θ component. $\qquad\square$

We now consider the Hamiltonian

$$H_\epsilon^s(\varphi, I, \tau) = H^s(\varphi, I) + \sqrt{\epsilon} P(\varphi, I, \tau), \quad \varphi \in \mathbb{T}^2, I \in \mathbb{R}^2, \tau \in \mathbb{T}_{\sqrt{\epsilon}}, \tag{11.2}$$

where $\|P\|_{C^2} \leq 1$.

Proposition 11.3. *Given any $\kappa > 0$, $0 < e < \bar{E}$ there is $\epsilon_0 > 0$ depending on H^s, e, κ, such that the following hold. For all $E_0 \in [e, \bar{E}]$, there is $E_1 < E_0 < E_2$, such that for all $0 < \epsilon < \epsilon_0$, there is an embedding*

$$\chi_\epsilon(x, y, \tau) : \mathbb{T} \times (E_1, E_2) \times \mathbb{T}_{\sqrt{\epsilon}} \to \mathbb{T}^2 \times \mathbb{R}^2 \times \mathbb{T}_{\sqrt{\epsilon}}, \quad \pi_\tau \chi_\epsilon = id$$

such that $\mathcal{C}_\epsilon$ is a weakly invariant normally hyperbolic cylinder for the Hamiltonian flow of H^s_ϵ (see (11.2)). Moreover, we have

$$\|\chi_\epsilon(\varphi, I, \tau) - \chi_0(\varphi, I)\|_{C^1} < \kappa,$$

and $\mathcal{C}_\epsilon$ contains all the invariant sets in

$$V = B_\kappa(\mathcal{C}_0) \times \mathbb{T}_{\sqrt{\epsilon}}.$$

Proof. Since $\|\sqrt{\epsilon}P\|_{C^2} \leq \sqrt{\epsilon}$, this is a *regular perturbation* of the vector field, compared to the *singular perturbation* we see in Chapter 9. The proof follows from standard theory, see for example [70, 31]. $\qquad \square$

We will denote by $\chi^0_\epsilon(x, y) = \chi_\epsilon(\varphi, I, 0)$ the zero-section of the embedding, and

$$\chi^0_\epsilon \left(\mathbb{T} \times (E_1, E_2) \right) = \mathcal{C}^0_\epsilon,$$

which is invariant under the time-$\sqrt{\epsilon}$ map $\phi = \phi^{\sqrt{\epsilon}}_{H^s_\epsilon}$. The following Lipschitz estimate is a crucial part of establishing Aubry-Mather type.

Proposition 11.4. *Given $R > 0$, there is $C > 0$ depending only on $\|\partial^2 K\|$, $\|H^s\|_{C^2}$, and R, such that if u is any weak KAM solution of H^s_ϵ at cohomology c with $\|c\| \leq R$, then for any*

$$(\varphi_1, I_1), \quad (\varphi_2, I_2) \in \tilde{\mathcal{I}}(c, u) = \bigcap_{k \geq \epsilon^{-1}} \phi^{-k} \left(\mathcal{G}_{c,u} \right),$$

we have

$$|H^s(\varphi_1, I_1) - H^s(\varphi_2, I_2)| \leq C\epsilon^{\frac{1}{4}} \|\varphi_1 - \varphi_2\|. \tag{11.3}$$

In particular, (11.3) holds on the sets $\tilde{\mathcal{I}}(c, u) = \bigcap_{n \in \mathbb{N}} \phi^{-n} \left(\mathcal{G}_{c,u} \right)$ and $\tilde{\mathcal{A}}^0_{H^s_\epsilon}$.

Proof. We apply Theorem 7.10 to the system $H^s_\epsilon = H^s + \sqrt{\epsilon}P$ and a continuous time weak KAM solution $w(x, t)$. We then obtain a $O(\sqrt{\epsilon^{\frac{1}{2}}}) = O(\epsilon^{\frac{1}{4}})$ Lipschitz estimate (note that the small parameter is $\sqrt{\epsilon}$), on the set $\phi^{-t}_{H^s_\epsilon}\mathcal{G}_{c,w}$, as long as $t > 1/\sqrt{\epsilon}$. Note that the corresponding discrete dynamics is $\phi = \phi^{\sqrt{\epsilon}}_{H^s_\epsilon}$, and therefore we need to take $k > 1/\epsilon$ to have $t = k\sqrt{\epsilon} > 1/\sqrt{\epsilon}$. $\qquad \square$

Proof of Theorem 11.1. Part (1): Let $c_0 = \bar{c}_h(E_0)$, and $\gamma^{E_0}_h$ is the unique shortest curve, we show (H^ϵ_s, c_0) is of Aubry-Mather type. By Proposition 11.3 and the semi-continuity of the Mañé set, for sufficiently small ϵ and σ, $\tilde{\mathcal{N}}_{H^s_\epsilon}(c_0) \subset B_\kappa(\mathcal{C}_0) \times \mathbb{T}_{\sqrt{\epsilon}}$ for all for $c \in \overline{B_\sigma(c_0)}$. As a result, $\tilde{\mathcal{A}}_{H^s_\epsilon}(c) \subset \tilde{\mathcal{A}}_{H^s_\epsilon}(c) \subset \mathcal{C}_\epsilon$.

We now prove the graph property. Suppose

$$(x_1, y_1), \quad (x_2, y_2) \quad \in (\chi^0_\epsilon)^{-1}(\tilde{\mathcal{A}}^0_{H^s_\epsilon}(c))$$

where $c \in \overline{B_\sigma(c_0)}$. Observe that $H^s \circ \chi_0(x, y) = y$. Then due to Proposition 11.4,

$$\|H^s \circ \chi^0_\epsilon(x, y) - y\|_{C^1} = \|H^s \circ \chi^0_\epsilon - H^s \circ \chi_0\| \leq \|H^s\|_{C^2}\|\chi^0_\epsilon - \chi_0\|_{C^1} \leq C\kappa,$$

where C will be used to denote a generic constant. We have

$$\|y_2 - y_1\| \leq \|H^s \circ \chi^0_\epsilon(x_2, y_2) - H^s \circ \chi^0_\epsilon(x_1, y_1)\| + C\kappa\|(x_2 - x_1, y_2 - y_1)\|$$
$$\leq C\epsilon^{\frac{1}{4}}\|\pi_\varphi \circ \chi^0_\epsilon(x_2, y_2) - \pi_\varphi \circ \chi^0_\epsilon(x_1, y_1)\| + C\kappa\|(x_2 - x_1, y_2 - y_1)\|$$
$$\leq C(\kappa + \epsilon^{\frac{1}{4}})\|(x_2 - x_1, y_2 - y_1)\|.$$

(11.4)

For ϵ, κ small enough, we have $\|y_2 - y_1\| \leq 2C(\kappa + \epsilon^{\frac{1}{4}})\|x_2 - x_1\|$.

Finally, assume that $(\chi^0_\epsilon)^{-1}\tilde{\mathcal{A}}^0_{H^s_\epsilon}(c)$ projects onto the x component. We will show that $W^u(\tilde{\mathcal{A}}^0_{H^s_\epsilon}(c))$ is a graph over the θ component. According to Lemma 11.2, $W^u(\tilde{\mathcal{A}}^0_{H^s_\epsilon}(c))$ is a smooth graph over θ component, and the projection of the unstable bundle is normal to γ^E_h. Let us note that (11.4) implies $\tilde{\mathcal{A}}^0_{H^s_\epsilon}(c)$ is close to $\tilde{\mathcal{A}}^0_{H^s_\epsilon}(c)$ in Lipschitz norm, and the unstable bundle depends smoothly on perturbation. Therefore the unstable bundle of $\tilde{\mathcal{A}}^0_{H^s_\epsilon}(c)$ is also transversal to $\mathcal{A}_{H^s_\epsilon}(c)$.

We still need to prove that $W^u(\tilde{\mathcal{A}}^0_{H^s_\epsilon}(c))$ is a Lipschitz graph over the θ component near the Aubry set $\mathcal{A}^0_{H^s_\epsilon}(c)$, if the Aubry set is an invariant circle (see Definition 8.1, 2b). Letting $\mathcal{G}$ denote the pseudograph associated to the unique weak KAM solution of H^s_ϵ at c, we will show that $\mathcal{G}$ and $W^u(\tilde{\mathcal{A}}^0_{H^s_\epsilon}(c))$ coincide on some neighborhood of $\mathcal{A}^0_{H^s_\epsilon}(c)$.

Let us extend the coordinates $\chi^0(x, y)$ to C^1 coordinates $\Lambda(x, y, s, u)$ in a neighborhood U of $\mathcal{C}_0$ such that the s coordinate axis is transversal to the unstable bundle, and the u coordinate axis is transversal to the stable bundle. We similarly extend $\chi^\epsilon_\epsilon(x, y)$ to $\Lambda^\epsilon(x, y, s, u)$, and we can do this keeping the estimate $\|\Lambda - \Lambda^\epsilon\|_{C^1} < C\kappa$ with the norm defined on $U(\mathcal{C})$.

Let $z \in \mathcal{G} \cap U$, then $\phi^{-k}(z)$ accumulates to $\mathcal{A}^0_{H^s_\epsilon}(c) \subset \mathcal{C}_\epsilon$ (ϕ is the time-$\sqrt{\epsilon}$ map of H^ϵ_s). It follows that there exists $\xi \in \mathcal{C}$ such that $z \in W^u(\xi)$. We claim that $\xi \in \tilde{\mathcal{A}}^0_{H^s_\epsilon}(c)$ (which implies $z \in W^u(\tilde{\mathcal{A}}^0_{H^s_\epsilon}(c))$). Assume this does not hold. For a sufficiently large k, due to normal hyperbolicity, there exist $\mu > 1$ and $\lambda < \mu$ such that

$$\|\phi^{-k}z - \phi^{-k}\xi\| < C\mu^{-k}, \quad \text{dist}(\phi^{-k}\xi, \tilde{\mathcal{A}}^0_{H^s_\epsilon}(c)) > C^{-1}\lambda^{-k}.$$

Let ζ_k denote the unique point in $\widetilde{\mathcal{A}}^0_{H^s_\epsilon}(c)$ such that ζ_k and $\phi^{-k}\xi$ has the same x component in the local coordinates (this is possible because $\widetilde{\mathcal{A}}^0_{H^s_\epsilon}(c)$ is a Lipschitz graph over x component). Choosing $k > 1/\epsilon$, we have:

- $\|\pi_{(x,y)}(\phi^{-k}\xi - \zeta_k)\| \geq \text{dist}(\phi^{-k}z, \widetilde{\mathcal{A}}^0_{H^s_\epsilon}(c)) > C^{-1}\lambda^{-k}$.
- $\|\pi_y(\phi^{-k}z - \zeta_k)\| \leq 2C(\kappa + \epsilon^{\frac{1}{4}})\|\pi_x(\phi^{-k}z - \zeta_k)\| \leq \|\pi_x(\phi^{-k}z - \zeta_k)\|$. Since both $\phi^{-k}z$ and ζ_k are contained in $\bigcap_{j \geq \epsilon^{-1}} \phi^{-j}\mathcal{G}$, Proposition 11.4 and (11.4) apply.
- $\|\pi_x(\phi^{-k}z - \zeta_k)\| = \|\pi_x(\phi^{-k}z - \phi^{-k}\xi)\| < C\mu^{-k}$.

Combining all the inequalities, we get

$$C^{-1}\lambda^{-k} \leq \|\pi_{(x,y)}(\phi^{-k}\xi - \zeta_k)\| \leq \|\pi_{(x,y)}(\phi^{-k}z - \zeta_k)\| + C\mu^{-k}$$
$$\leq 2\|\pi_x(\phi^{-k}z - \zeta_k)\| + C\mu^{-k} \leq 3C\mu^{-k},$$

but this is a contradiction when k is large enough. We have proven that $\mathcal{G} \cap U$ is contained in $W^u\widetilde{\mathcal{A}}^0_{H^s_\epsilon}(c)$. By Proposition 6.14, $\phi^{-1}\mathcal{G}$ is a Lipschitz graph. Taking U smaller, we have $\phi^{-1}\widetilde{\mathcal{G}} \cap U = W^u\widetilde{\mathcal{A}}^0_{H^s_\epsilon}(c) \cap U$ is also a Lipschitz graph. This proves 2b) of Definition 8.1.

Part (2): Suppose $c_0 = \bar{c}_h(E_0)$ is such that $\mathcal{A}_{H^s}(c) = \gamma_h^{E_0} \cup \bar{\gamma}_h^{E_0}$ are two shortest curves. Let V_1, V_2 be neighborhoods of $\gamma^{E_0}, \bar{\gamma}^{E_0}$ such that $\pi_\varphi \eta^E \subset V_1$, $\pi_\varphi \bar{\eta}^E \subset V_2$ for all $E \in (E_0 - \delta, E_0 + \delta)$, and let

$$f : \mathbb{T}^n \to [0,1], \quad f|_{V_1} = 0, \quad f|_{V_2} = 1$$

be a smooth bump function. Then for

$$H^1 = H^s - f, \quad H^2 = H^s - (1 - f),$$

$\widetilde{\mathcal{A}}_{H^1}(c_0) = \gamma^{E_0}$ and $\widetilde{\mathcal{A}}_{H^2}(c_0) = \bar{\gamma}^{E_0}$. We then apply part (1) to obtain that for each $i = 1, 2$, $H^i + \sqrt{\epsilon}P, c_0$ is of Aubry-Mather type. This implies H^s_ϵ, c_0 is of bifurcation Aubry-Mather type (see Definition 8.3). $\square$

11.2 SIMPLE NON-CRITICAL CASE

Suppose h is a simple non-critical homology, which means that for the energy $E = 0$, there is a unique shortest curve γ_h^0 in the homology h corresponding to a hyperbolic periodic orbit η_h^0 of the Hamiltonian system. In this case, however, we have, for $c_0 = \bar{c}_h(0)$

$$\mathcal{A}_{H^s}(c_0) = \gamma_h^0 \cup O,$$

where O is the origin (which is where U attains its minimum). If we consider V_1, V_2 disjoint open sets containing γ_h^0 and 0 respectively, and define the local

Aubry sets $\mathcal{A}_{H^1}(c)$ and $\mathcal{A}_{H^2}(c)$, it follows that $\mathcal{A}_{H^1}(c) = \gamma_h^0$ and $\mathcal{A}_{H^2}(c) = O$.

Theorem 11.5. *Suppose h is simple non-critical, and let $c_0 = \bar{c}_h(0)$. Then there are $\epsilon_0, \delta > 0$ depending only on K, U such that for $0 < \epsilon < \epsilon_0$ and $U' \in \mathcal{V}_\delta(U)$, for*

$$H_\epsilon^s = K(I) - U'(\varphi) + \sqrt{\epsilon}P, \quad \|P\|_{C^2} \leq 1,$$

the pair (H_ϵ^s, c_0) is of asymmetric bifurcation type.

Proof. The fact that H^1, c_0 is of Aubry-Mather type follows the same proof as Theorem 11.1. On the other hand, $\tilde{\mathcal{A}}_{H^s}(c_0)$ is a hyperbolic periodic orbit, which is robust under perturbation. Therefore Definition 8.5 is satisfied.   □

11.3   SIMPLE CRITICAL CASE

11.3.1   Proof of Aubry-Mather type using local coordinates

Theorem 11.6. *Suppose $h \in H_1(\mathbb{T}^2, \mathbb{Z})$ is a simple homology for H^s, and consider the cohomology class $c_0 = \bar{c}_h(0)$. Then there exist $\epsilon_0, e, \delta > 0$ depending only on H^s, h such that for each $0 < \epsilon < \epsilon_0$, $U' \in \mathcal{V}_\delta(U)$ and $c \in B_e(c_0)$, the pair (H_ϵ^s, c) is of Aubry-Mather type.*
Moreover, the same holds for $(H^s, \lambda c)$ for all $0 \leq \lambda \leq 1$.

We have the following (See Chapter 10):

- $\eta_h^0 = \tilde{\mathcal{A}}_{H^s}(c_0)$ contains the hyperbolic fixed point $(0,0)$, and is a homoclinic orbit to $(0,0)$.
- $(0,0)$ admits eigenvectors $-\lambda_2 < -\lambda_1 < \lambda_1 < \lambda_2$. Let $v_1^{s/u}$ and $v_2^{s/u}$ denote the eigendirections of the eigenvalues $\pm\lambda_1, \pm\lambda_2$. Let $\mathrm{Inv}(\varphi, I) = (\varphi, -I)$ denote the involution of the Hamiltonian system. Since the flow is time-reversible, we have $\mathrm{Inv}(v_i^s) = \pm v_i^u$, $i = 1, 2$. Without loss of generality, we assume $\mathrm{Inv}(v_i^s) = v_i^u$. As a result, there exist $v_i, w_i \in \mathbb{R}^2$ such that

$$v_i^s = (v_i, w_i), \quad v_i^u = (v_i, -w_i), \quad i = 1, 2. \tag{11.5}$$

- There is a C^1 normally hyperbolic invariant manifold $\mathcal{M}$ containing η_h^0. In particular, $\mathcal{M}$ must contain $(0,0)$ and it's tangent to the plane $\mathrm{Span}\{v_1^s, v_1^u\} = \mathbb{R}v_1 \oplus \mathbb{R}w_1 \subset \mathbb{R}^2 \times \mathbb{R}^2$ at $(0,0)$.
  The projection $\pi_\varphi \eta_h^0 = \gamma_h^0$ then is a C^1 curve in $\mathbb{T}^2$, since $0 \in \mathbb{T}^2$ is the only possible discontinuity of the tangent direction, but at 0 the curve is tangent to $v_1 = \pi_\varphi v_1^s = \pi_\varphi v_1^u$. Let $\pi_{v_1} : \mathbb{R}^4 \to \mathbb{R}$ and $\pi_{w_1} : \mathbb{R}^4 \to \mathbb{R}$ be the orthogonal projections to v_1, w_1 directions.

We also need the following analog of Lemma 11.2.

Lemma 11.7. *$W^u(\eta_h^0)$ is a Lipschitz graph over the θ component on a neigh-*

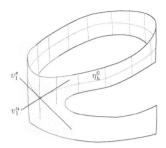

Figure 11.1: Coordinates near the homoclinic orbit: vertical lines indicate the level curves of the x coordinate.

borhood of γ_h^0.

Proof. In the proof of Lemma 11.2, the Lipschitz constant of the Green bundles is uniform over all energy E. The lemma follows by taking the limit $E \to 0$ in the space of Lipschitz graphs. $\square$

We require a suitable parametrization of the cylinder $\mathcal{C}(H^s)$ near the homoclinic η_h^0. An illustration of the parametrization is given in Figure 11.1.

Proposition 11.8. *For each $\kappa > 0$ there exist $\delta_1, \delta_2 > 0$, and a smooth embedding*

$$\chi_0 = \chi_0(x, y): \quad \mathbb{T} \times (-\delta_2, \delta_2) \to \mathcal{M},$$

such that the following hold:

1. $\mathcal{C}(H^s) := \chi_0(\mathbb{T} \times (-\delta_2, \delta_2)) \supset \eta_h^0$. *(We use the notation $\mathcal{C}$, since the image of χ_0 is a cylinder.)*
2. *The cylinder $\mathcal{C}(H^s)$ is symplectic.*
3. *χ_0 is "almost vertical" near $x = 0$, namely $\pi_\varphi \circ \chi_0(x, y)$ is κ-Lipschitz in y for all $|x| < \delta_1$.*
4. *The vertical coordinate is given by the energy function away from $x = 0$, namely for all $|x| \geq \delta_1/2$ we have $H^s(\chi_0(x, y)) = y$.*

We now prove Theorem 11.6 assuming Proposition 11.8.

Proof of Theorem 11.6. Let us consider the system

$$H_\epsilon^s = K(I) + U'(\varphi) + \sqrt{\epsilon} P.$$

where $\|P\|_{C^2} \leq 1$ and $\|U - U'\|_{C^2} \leq \delta$. Then for ϵ_0, δ small enough depending only on K, U, standard perturbation theory implies H_ϵ^s admits a normally hyperbolic weakly invariant manifold $\mathcal{C}(H_\epsilon^s)$, let us denote by $\mathcal{C}_\epsilon$ its zero section. Then $\mathcal{C}_\epsilon$ is invariant under the time-$\sqrt{\epsilon}$ map $\Phi = \Phi_{H_\epsilon^s}^{\sqrt{\epsilon}}$, and it admits a parametrization $\chi_\epsilon : \mathbb{T} \times (-\delta_2, \delta_2) \to \mathcal{C}_\epsilon$, and $\|\chi_\epsilon - \chi_0\|_{C^1} = o(1)$ as $\epsilon, \delta \to 0$. The cylinder is symplectic, since symplecticity is open under perturbations. Moreover, for $e > 0$ small enough, the set $\widetilde{\mathcal{N}}_{H^s}(c) \subset \mathcal{C}(H^s)$ for all $c \in B_e(c_0)$, and since $\widetilde{\mathcal{N}}$ is upper semi-continuous under Hamiltonian pertubations, $\widetilde{\mathcal{N}}_{H^s_\epsilon}(c)$ is close to $\mathcal{C}(H^s_\epsilon)$ for small ϵ. Since $\mathcal{C}(H^s_\epsilon)$ contains all the invariant sets in its neighborhood, we conclude that $\widetilde{\mathcal{A}}_{H^s_\epsilon}(c) \subset \widetilde{\mathcal{N}}_{H^s_\epsilon}(c) \subset \mathcal{C}(H^s_\epsilon)$.

We now show $\chi_\epsilon^{-1} \widetilde{\mathcal{A}}_{H^s_\epsilon}(c)$ is a Lipschitz graph for $c \in B_e(c_0)$. Let $(\varphi_i, I_i) = \chi_\epsilon(x_i, y_i)$, $i = 1, 2$ be two points in $\widetilde{\mathcal{A}}_{H^s_\epsilon}(c)$.

The proof consists of two cases. In the first case we use the almost verticality of the cylinder, and the idea is similar to the proof of Theorem 9.3. In the second case we use the strong Lipschitz estimate for the energy H^s, and the idea is similar to the proof of Theorem 11.1.

Case 1. $|x_1|, |x_2| < \delta_1$. In this case, we apply the a priori Lipschitz estimates for the Aubry sets: there is $C > 0$ depending only $\partial^2 K$, $\|H^s\|_{C^2}$ such that

$$\|I_2 - I_1\| \leq C\|\varphi_2 - \varphi_1\|.$$

Let κ be as in Proposition 11.8, and let ϵ_0 be small enough such that $\pi_\varphi \chi_\epsilon(x, y)$ is 2κ-Lipschitz in y. Since χ_0 is an embedding, there is $C > 1$ depending only on H^s such that for all ϵ_0 small enough,

$$\|\varphi_2 - \varphi_1\| + \|I_2 - I_1\| \geq C^{-1} \left(\|x_2 - x_1\| + \|y_2 - y_1\| \right).$$

Let $(\varphi_3, I_3) = \chi_\epsilon(x_2, y_1)$ then

$$\|\varphi_3 - \varphi_2\| \leq \|\chi_\epsilon(x_2, y_1) - \chi_\epsilon(x_2, y_2)\| \leq 2\kappa\|y_2 - y_1\|,$$

$$\|\varphi_3 - \varphi_1\| \leq \|\chi_\epsilon(x_2, y_1) - \chi_\epsilon(x_1, y_1)\| \leq C\|x_2 - x_1\|.$$

Combining all estimates, we get

$$C^{-1} \left(\|x_2 - x_1\| + \|y_2 - y_1\| \right) \leq (1 + C)\|\varphi_2 - \varphi_1\|$$
$$\leq (1 + C) \left(2\kappa\|y_2 - y_1\| + C\|x_2 - x_1\| \right),$$

or

$$\left(C^{-1} - 2(1 + C)\kappa \right) \|y_2 - y_1\| \leq \left(C^{-1} + (1 + C)C \right) \|x_2 - x_1\|,$$

which is what we need if $\kappa < 1/(4C(1 + C))$.

Case 2. $|x_1|, |x_2| > \delta_1/2$. In this case we apply Proposition 11.4, to get

$$\|H^s(\varphi_2, I_2) - H^s(\varphi_1, I_1)\| \leq C\epsilon^{\frac{1}{4}}\|\varphi_2 - \varphi_1\|.$$

Assume that ϵ is small enough such that $\|\chi_\epsilon - \chi_0\|_{C^1} < \kappa$. Then a computation identical to (11.4) implies $\|y_2 - y_1\| \leq 2C(\kappa + \epsilon^{\frac{1}{4}})\|x_2 - x_1\|$ if ϵ, κ is small enough depending only on C.

We obtain the Lipschitz property of $\chi_\epsilon^{-1} \widetilde{\mathcal{A}}^0_{H^s_\epsilon}(c)$ after combining the two cases.

We now check that the unstable bundle E^u is uniformly transverse to the projection $\mathcal{A}^0_{H^s_\epsilon}(c)$. In case 1, the almost verticality of the cylinder implies $\mathcal{A}^0_{H^s_\epsilon}(c)$ differs from $\mathcal{A}_{H^s}(c_0) = \gamma^0_h$ by $O(\kappa)$ in Lipschitz norm. Given that the tangent vector of γ^0_h is close to the weak directions $v^{s/u}_1$ in case 1, while the projection of E^u is close to $v^{s/u}_2$, the desired transversality holds. In case 2, similar to the proof of Theorem 11.1, $\widetilde{\mathcal{A}}_{H^s}(c)$ is also close to $\widetilde{\mathcal{A}}_{H^s}(c_0)$ in Lipschitz norm (for a different reason, i.e (11.4)). The claim follows, similar to the proof of Theorem 11.1, using Lemma 11.7.

Finally, we show $W^u(\widetilde{\mathcal{A}}^0_{H^s_\epsilon}(c))$ is a graph over φ when $\chi^{-1}_\epsilon \widetilde{\mathcal{A}}^0_{H^s_\epsilon}(c)$ projects onto the x component. Similar to the proof of Theorem 11.1, it suffices to show that if $\mathcal{G}$ is a weak KAM solution of H^s_ϵ at cohomology c, there is a neighbor U of $\mathcal{G}$ such that any $z \in \mathcal{U} \cup U$ and $\zeta \in \mathcal{C}$ such that $z \in W^u(\zeta)$, we can conclude that $z \in \widetilde{\mathcal{A}}^0_{H^s_\epsilon}(c)$. Arguing by contradiction as before, we suppose $\zeta \notin \widetilde{\mathcal{A}}^0_{H^s_\epsilon}(c)$ and let ξ_k be the unique point on $\widetilde{\mathcal{A}}^0_{H^s_\epsilon}(c)$ with the same x component as $\phi^{-k}\zeta$. Just as before, we can extend the coordinate χ_ϵ to a tubular neighborhood of $\mathcal{C}$. We obtain in the same fashion

- $\|\pi_{(x,y)}(\phi^{-k}\xi - \zeta_k)\| \geq \text{dist}(\phi^{-k}z, \widetilde{\mathcal{A}}^0_{H^s_\epsilon}(c)) > C^{-1}\lambda^{-k}$.
- $\|\pi_x(\phi^{-k}z - \zeta_k)\| = \|\pi_x(\phi^{-k}z - \phi^{-k}\xi)\| < C\mu^{-k}$.

The remaining inequality to prove is $\|\pi_y(\phi^{-k}z - \zeta_k)\| \leq C\|\pi_x(\phi^{-k}z - \zeta_k)\|$, which we split into Case 1 and 2 as before. Case 2 is identical to Theorem 11.1. Case 1 follows the same computation, with the observation that we can define the extension in such a way that $\pi_\varphi \Lambda(x, y, s, u)$ is still 2κ-Lipschitz in y. $\quad\square$

11.3.2   Construction of the local coordinates

This is done separately near the hyperbolic fixed point (local) and away from it (global). Furthermore, the local coordinate requires a preliminary step.

Lemma 11.9. *For each $\kappa > 0$ there is $\delta > 0$, and a smooth embedding*

$$\chi_{\text{pre}} = \chi_{\text{pre}}(x, y) : (-\delta, \delta) \times (-\delta, \delta) \to \mathcal{M} \cap B_{2\delta}(0, 0) \subset \mathbb{T}^2 \times \mathbb{R}^2,$$

satisfying

$$\|\chi_{\text{pre}} \circ (\pi_{v_1}, \pi_{w_1}) - \text{id}\|_{C^1} < \kappa. \tag{11.6}$$

There is $0 < \delta_1 < \delta$ such that the curve $\chi^{-1}_{\text{pre}}\eta^0_h \cap \pi^{-1}_x\{(-\delta_1, \delta_1)\}$ is given by a Lipschitz graph $\{(x, g(x)) : x \in (-\delta_1, \delta_1)\}$.

Proof. The existence of the local coordinate follows from $T_{(0,0)}\mathcal{M} = \mathbb{R}v_1 \oplus \mathbb{R}w_1$, the fact that $\mathcal{M}$ is C^1, and the implicit function theorem. Since $\gamma^0_h = \pi_\varphi \eta^0_h$ is a C^1 curve tangent to v_1 at $0 \in \mathbb{T}^2$, γ^0_h can be reparametrized using its projection to the v_1 direction. Since η^0_h is a Lipschitz graph over γ^0_h, the second claim follows.   □

Lemma 11.10. *For each $\kappa > 0$ there is $\delta_1 > \delta_2 > 0$, a smooth embedding*

$$\chi_{\mathrm{loc}} = \chi_{\mathrm{loc}}(x,y) : \quad (-\delta_1, \delta_1) \times (-\delta_1, \delta_1) \to \mathcal{M} \cap B_{2\delta}(0,0),$$

a neighborhood V of the local homoclinic $\eta_h^0 \cap \pi_x^{-1}\{(-\delta_1, \delta_1)\}$ on which the cylinder is "almost vertical" in the sense that there is $C > 0$ such that

$$\pi_\varphi \chi_{\mathrm{loc}}(x,y) \quad \text{is} \quad C\kappa - Lipschitz \text{ in } y.$$

The pull back $\chi_{\mathrm{loc}}^{-1}\eta_h^0 \cap \pi_x^{-1}\{(-\delta_1, \delta_1)\}$ is a Lipschitz graph over x, and in addition,

$$H^s(\chi_{\mathrm{loc}}(x,y)) = y, \quad \text{for all} \quad \delta_1/2 < |x| < \delta_1, |y| < \delta_2. \tag{11.7}$$

The manifold $\chi_{\mathrm{loc}}(-\delta_1, \delta_1) \times (-\delta_1, \delta_1))$ is symplectic.

Proof. First we show the image of χ_{pre} is symplectic. To see this, note that $H^s(\varphi, I) = \frac{1}{2}AI \cdot I - \frac{1}{2}B\varphi \cdot \varphi + O_3(I, \varphi)$, where $A = \partial_{II}^2 K$ and $B = \partial_{\varphi\varphi}^2 U(0)$ are both positive definite. Moreover, we have $\lambda_1 w_1 = Av_1$ and $\lambda_1 v_1 = Bw_1$, where v_1, w_1 are the vectors from (11.5). From (11.6), we have

$$\partial_x \chi_{\mathrm{pre}}(x,y) = v_1 + O(\kappa), \quad \partial_y \chi_{\mathrm{pre}}(x,y) = w_1 + O(\kappa),$$

Let ω be the standard symplectic form; it follows that

$$\omega(\partial_x \chi_{\mathrm{pre}}, \partial_y \chi_{\mathrm{pre}}) = \omega((v_1, 0), (0, w_1)) + O(\kappa) = \lambda_1 v_1 \cdot Av_1 + O(\kappa)$$

is uniformly bounded away from 0 if κ is small enough.

Let $\delta, \delta_1, \chi_{\mathrm{pre}}$ and g be from Lemma 11.9. We claim that after possibly shrinking δ_1, there is $\delta_2 > 0$ and $C > 1$ such that for all $(x, g(x) + y) \in \mathbb{R}^2$, $|y| < \delta_2, \delta_1/2 < |x| < \delta_1$ we have

$$C\delta_1 > |\partial_y(H^s \circ \chi_{\mathrm{pre}}(x, g(x) + y))| > C^{-1}\delta_1 > 0.$$

We will only prove this claim for the case $\delta_1/2 < x < \delta$, as the other half is symmetric.

Since η_h^0 is tangent to the stable/unstable vectors $v_1^{s/u}$, we have

$$\chi_{\mathrm{pre}}(x, g(x)) = xv_1^u + O(x^2) = x(v_1, w_1) + O(x^2),$$
$$\chi_{\mathrm{pre}}(x, g(x) + y) = x(v_1, w_1) + O(x^2) + O(y), \quad x > 0.$$

We have

$$\partial_y \left(H^s \circ \chi_{\text{pre}}(x, g(x) + y) \right) = \left(x(Bv_1, Aw_1) + O(x^2) + O(y) \right) \cdot (w_1 + O(\kappa))$$
$$= xAw_1 \cdot w_1 + O(x^2) + O(y) + O(\kappa(|x| + |y|))$$
$$\geq 4C^{-1}\delta_1 + O(x^2) + O(y) + O(\kappa(|x| + |y|)) > C^{-1}\delta_1,$$

if $8C^{-1} = \|Aw_1 \cdot w_1\|$, $\delta_1/2 < x < \delta_1$, $\delta_1 < C^{-1}$, $|y| < \delta_2 < C^{-1}\delta_1$, and $\kappa < C^{-1}/2$. The upper bound can be obtained similarly. This proves our claim.

We consider the function

$$F = F(x, y) = \frac{\delta_1^{-1}}{\partial_y \left(H^s \circ \chi_{\text{pre}}(x, g(x) + y) \right)}, \qquad \delta_1/2 < |x| < \delta_1, \ |y| < \delta_2.$$

and for each fixed x, let $Y_F(x, y)$ denote the solution to the ODE

$$\frac{d}{dy} Y_F = F(x, y), \qquad Y_F(0) = g(x),$$

then $\partial_y H^s \circ \chi_{\text{pre}}(x, Y_F(x, y)) = 1$, and therefore $H^s \circ \chi_{\text{pre}}(x, Y_F(x, y)) = y$.

Finally, let us define the vector field $G(x, y)$ via

$$G(x, y) = \begin{cases} (0, F(x, y)), & \delta_1/2 < |x| < \delta_1, \ |y| < \delta_2; \\ (0, 1), & |x| < \delta_1/4, \text{ or } |y| > 2\delta_2; \end{cases}$$

and smoothly interpolated (keeping the first coordinate 0) in between. Since $C^{-1} < |F_1| < C$, this can be done keeping $\|G\|_{C^1} = O(\delta_1^{-1})$. Let $g_1(x)$ be a mollified version of g such that $g_1(x) = g(x)$ for all $|x| \geq \delta_1/4$, and $|g_1(x) - g(x)| \leq \delta_1/2$. Finally define

$$\chi_{\text{loc}}(x, y) = \chi_{\text{pre}} \circ \Phi_G^y(x, g_1(x)),$$

where $\Phi_G(x_0, y_0)$ is the time-y-flow of $G(x, y)$. The modification g_1 is to ensure the coordinate system is smooth. See Figure 11.2.

We have:

- $\chi_{\text{loc}}(x, y) = \chi_{\text{pre}}(x, g_1(x) + y)$ when $|x| < \delta_1/4$ and $|y| > 2\delta_1$.
- $H^s \circ \chi_{\text{loc}}(x, y) = H^s \circ \chi_{\text{pre}}(x, Y_F(x, y)) = y$ when $\delta_1/2 < |x| < \delta_1$ and $|y| < \delta_2$.
- The function χ_{pre} is κ-Lipschitz in y (see (11.6)). Therefore (11.7) holds when $|x| < \delta_1/4$ and $|y| > 2\delta_1$.
  Moreover, given that the flow Φ_G^y is vertical, we have $\pi_x \Phi_G^y(x_0, y_0) = x_0$ and $|\partial_{y_0} \Phi_G^y| \leq 1 + |y| \cdot \|G\|_{C^1} = 1 + O(\delta_1^{-1}|y|)$. Therefore

$$|\partial_y(\pi_\varphi \chi_{\text{pre}}) \circ \Phi_G^y(x, g_1(x))| \leq \|\partial_y(\pi_\varphi \chi_{\text{pre}})\| \cdot \left(1 + O(\delta_1^{-1}|y|) \right) \leq 2\kappa$$

if $|y| < \delta_2$ is sufficiently small. As a result, (11.7) holds on the set $V := \{|y| < \delta_2\}$, which is the gray area in Figure 11.2.

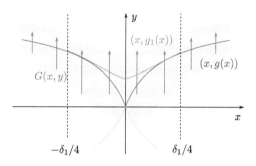

Figure 11.2: Construction of local coordinate.

- The curve $\eta_h^0 = \{\chi_{\mathrm{pre}}(x, g(x)) : |x| < \delta_1\}$ coincides with $\chi_{\mathrm{loc}}(x, 0)$ when $|x| \geq \delta_1/4$ and coincides with $\chi_{\mathrm{loc}}(x, y - g_1(x) + g(x))$ when $|x| \leq \delta_1/4$, and therefore is a Lipschitz graph under χ_{loc}^{-1}.

$\square$

Proof of Proposition 11.8. Consider the χ_{loc} coordinate system as constructed. By construction, the sections $\Sigma_\pm = \chi_{\mathrm{loc}}(\{|y| \leq \delta_2, x = \pm\delta_1\})$ are transversal to the Hamiltonian flow, and therefore there is a Poincaré map $\Phi : \Sigma_+ \to \Sigma_-$. Moreover, we must have $\chi_{\mathrm{loc}}^{-1} \circ \Phi \circ \chi_{\mathrm{loc}}(\delta_1, y) = (-\delta_1, y)$, since the flow preserves energy. Let $T(y)$ denote the time it takes for $\phi_{H^s}^t$ to flow from $\chi_{\mathrm{loc}}(\delta_1, y)$ to $\chi_{\mathrm{loc}}(-\delta_1, y)$.

We now define

$$\chi(x, y) = \begin{cases} \chi_{\mathrm{loc}}(x, y), & 0 \leq x < \delta_1; \\ \chi_{\mathrm{loc}}(x - 1, y), & 1 - \delta_1 \leq x \leq 1; \\ \phi_{H^s}^{s(x,y)}(\delta_1, y), & s(x, y) = \frac{x - \delta_1}{1 - 2\delta_1} T(y), \quad \delta_1 \leq x \leq 1 - \delta_1. \end{cases}$$

Then $\chi(x, y)$ is an embedding $\mathbb{T} \times (-\delta_2, \delta_2) \to \mathbb{T}^2 \times \mathbb{R}^2$ satisfying item 1, 3, and 4 of Proposition 11.8.

It remains to prove (2), namely symplecticity. For $|x| \leq \delta_1$, this is covered in Proposition 11.8. For $|x| \geq \delta_1$, note that the tangent plane to the cylinder is spanned by two vector fields, one being X_{H^s} and the other is $\partial_y \chi$. We have $\omega(X_H, \partial_y \chi) = \nabla H \cdot \partial_y \chi = 1$ by construction, therefore the manifold is symplectic. $\square$

Chapter Twelve

Forcing equivalence between kissing cylinders

In this section we prove Theorem 5.9, i.e. the *jump mechanism*. We assume that h is a non-simple homology class which satisfies $h = n_1 h_1 + n_2 h_2$, where h_1, h_2 are simple homologies. They are associated with curves $\bar{c}_h(E)$ and $\bar{c}_{h_1}^\mu(E)$ in the corresponding channels, where we assume

$$\left\| \bar{c}_h(0) - \bar{c}_{h_1}^\mu(0) \right\| < \mu.$$

Our goal is to prove forcing equivalence of the cohomologies $\Phi_L^*(\bar{c}_h(E_1))$ and $\Phi_L^*(\bar{c}_{h_1}^\mu(E_2))$ for some $E_1, E_2 \in (e, e + \mu)$, where $e > 0$ is sufficiently small.

We will construct a variational problem which proves forcing equivalence for the original Hamiltonian H_ϵ using Definition 6.18. The proof consists of the following steps.

1. We construct a special variational problem for the slow mechanical system H^s. A solution of this variational problem is an orbit "jumping" from one homology class h to the other h_1. The same can be done with h and h_1 switched.
2. We then modify this variational problem for the fast time-periodic perturbation of H^s, i.e. for the perturbed slow system $H_\epsilon^s(\varphi^s, I^s, \tau) = K(I^s) - U(\varphi^s) + \sqrt{\epsilon} P(\varphi^s, I^s, \tau)$ with $\tau \in \sqrt{\epsilon}\, \mathbb{T}$. This is achieved by applying the perturbative results established in Chapter 7.

   Recall the original Hamiltonian system H_ϵ near a double resonance can be brought to a normal form $N_\epsilon = H_\epsilon \circ \Phi_\epsilon$ and this normal form, in turn, is related to the perturbed slow system through coordinate change and energy reduction (see Chapter 14). The variational problem for H_ϵ^s can then be converted to a variational problem for the original H_ϵ.
3. Using this variational problem we prove the forcing equilvalence between $\Phi_L^*(\bar{c}_h(E_1))$ and $\Phi_L^*(\bar{c}_{h_1}^\mu(E_2))$.

12.1 VARIATIONAL PROBLEM FOR THE SLOW MECHANICAL SYSTEM

The slow system is given by $H^s(\varphi, I) = K(I) - U(\varphi)$, where we assumed that the minimum of U is achieved at 0. Given $m \in \mathbb{T}^2$, $a > 0$ and a unit vector

$\omega \in \mathbb{R}^2$, define
$$S(m, a, \omega) = \{m + \lambda\omega : \lambda \in (-a, a)\}.$$

$S(m, a, \omega)$ is a line segment in $\mathbb{T}^2$ and we will refer to it as a *section* (see Figure 12.1).

Given $c_1, c_2 \in \mathbb{R}^2$, we say that c_1, c_2 have a non-degenerate connection for H^s, at the section $S(m, a, \omega) \in \mathbb{T}^2$, if the following conditions hold.

[N1A]
$$(c_1 - c_2) \perp S, \quad \alpha_H(c_1) = \alpha_H(c_2).$$

[N2A] There exists a relatively compact set $K \subset S$ such that for all $x \in \mathcal{A}_H(c_1)$ and $z \in \mathcal{A}_H(c_2)$, the minimum of the variational problem
$$\min_{y \in \overline{S}} \{h_{H^s, c_1}(x, y) + h_{H^s, c_2}(y, z)\}$$
is never achieved outside of K.

[N3A] Suppose the above minimum is achieved at y_0; let $p_1 - c_1$ be any super-differential of $h_{H^s, c_1}(x, \cdot)$ at y_0, and $-p_2 + c_2$ a super-differential of $h_{H^s, c_2}(\cdot, z)$ at y_0, then
$$\partial_p H^s(y_0, p_i) \cdot S^\perp, \quad i = 1, 2$$
have the same signs; here $S^\perp$ denotes a normal vector to S.

Remark 12.1. • The "A" in [N1A]–[N3A] stands for "autonomous."
• Conditions of the type [N1A]–[N3A] are common in variational construction of shadowing orbits, called the "no corners" conditions (see for example [15]). They imply that the minimizers of the $h_{H^s, c_1}(x, y)$ and $h_{H^s, c}(y, z)$ concatenate to a smooth trajectory of the Euler-Lagrange equation. We take advantage of this fact to prove the forcing equivalence between c_1 and c_2 using the defintion.
• In the sequel, the sections S chosen will always be an affine manifold without boundary of co-dimension 1, and hence the perpendicular direction $S^\perp$ is well defined. We will sometimes abbreviate the phrase "the minimum is never reached outside of a relatively compact subset K of S" as "has interior minimum."

Recall that the cohomology classes $\bar{c}_h(E)$, $\bar{c}_{h_1}^\mu(E)$ are chosen in the channels of h and h_1, i.e. $\mathcal{LF}(\lambda_h^E h)$ and $\mathcal{LF}(\lambda_{h_1}^E h_1)$. Since h is non-simple, Proposition 15.11 implies the channel will pinch to a point, and $\bar{c}_h(0)$ is uniquely chosen. The channel of h_1 has positive width at $E = 0$, and $\bar{c}_h(0)$ is at the boundary of this segment. In particular, since the channel for h_1 is parallel to $h_1^\perp$, we always have
$$\bar{c}_{h_1}^\mu(0) - \bar{c}_h(0) \perp h_1. \tag{12.1}$$

We will choose $\bar{c}_{h_1}(0)$ sufficiently close to $\bar{c}_h(0)$ according to the following proposition.

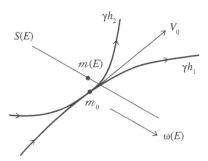

Figure 12.1: Jump from one cylinder to another in the same homology.

Proposition 12.2. *Suppose the slow mechanical system H^s satisfies conditions [A0]–[A4] Then there is $\mu_0 > 0$ depending only on H^s such that if $0 < \mu < \mu_0$, we have: There exists a section $S(0) = S(m(0), a(0), \omega(0))$, such that conditions [N1A]–[N3A] are satisfied for $\bar{c}_{h_1}(0)$, $\bar{c}_h(0)$ and section $S(0)$.*
Moreover, the same holds with $\bar{c}_{h_1}(0)$, $\bar{c}_h(0)$ switched.

Proof. We first show that [N1A]–[N3A] are satisfied when $c_1 = c_2 = \bar{c}_h(0)$, then use perturbation arguments. Note that [N1A] is trivially satisfied when $c_1 = c_2$.

Recall that $\mathcal{A}_{H^s}(\bar{c}_h(0)) = \gamma^0_{h_1} \cup \gamma^0_{h_2}$, and the curves $\gamma^0_{h_1}$ and $\gamma^0_{h_2}$ are tangent to a common direction at m_0, which we will call v_0. By the choice of h_1, v_0 is not parallel to h_1. We now choose $\omega(0) = \frac{h_1}{\|h_1\|}$, $m(0)$ sufficiently close to 0, and $a(0) = a > 0$, such that $S(m(0), \omega(0), a(0))$ intersects $\mathcal{A}_{H^s}(\bar{c}_h(0))$ transversally and is disjoint from 0 (see Figure 12.1).

The Aubry set $\tilde{\mathcal{A}}_{H^s}(\bar{c}_h(0))$ supports a unique minimal measure, and therefore has a unique static class. For any $x, z \in \mathcal{A}_{H^s}(\bar{c}_h(0))$, the function

$$h_{\bar{c}_h(0)}(x, \cdot) + h_{\bar{c}_h(0)}(\cdot, z)$$

reaches its global minimal at $\mathcal{N}_{H^s}(\bar{c}_h(0))$, which coincides with $\mathcal{A}_{H^s}(\bar{c}_h(0))$. Hence the minimum in

$$\min_{y \in S(0)} \left\{ h_{\bar{c}_h(0)}(x, y) + h_{\bar{c}_h(0)}(y, z) \right\} \tag{12.2}$$

is reached at $S(0) \cap \mathcal{A}_{H^s}(\bar{c}_h(0))$, which consists of finitely many points, and hence is compactly contained in $S(0)$. This implies [N2A] is satisfied for $c_1 = c_2 = \bar{c}_h(0)$ along the section $S(0)$.

Moreover, for any y_0 reaching the minimum in (12.2), according to the above analyis y_0 also reaches the global minimum of $h_{\bar{c}_h(0)}(x, \cdot) + h_{\bar{c}_h(0)}(\cdot, z)$. Then using Proposition 6.2, the associated super-gradients $p_1 - \bar{c}_h(0)$ and $-p_2 + \bar{c}_h(0)$ (in [N3A]) satisfy $p_1 = p_2$, and $\partial_p H^s(y_0, p_1) = \partial_p H^s(y_0, p_2)$ is the velocity of the unique backward minimizer at y_0. To show that [N3A] holds, we only need to show $\partial_p H^s(y_0, p_1) \neq 0$. This is the case because 0 is the only equilibrium in

$\mathcal{A}_{H^s}(\bar{c}_h(0))$, and $S(0)$ is disjoint from 0.

We now perturb c_1 away from $\bar{c}_h(0)$ while keeping $c_2 = \bar{c}_h(0)$. We choose $c_1 = \bar{c}^\mu_{h_1}(0)$ at the bottom of the h_1 channel, then (12.1) and $\omega(0) = h_1$ means [N1A] is still satisfied. Proposition 7.4 implies that [N2A] and [N3A] are both robust under the perturbation of c_1, and therefore [N1A]–[N3A] holds if $\|\bar{c}_{h_1}(0) - \bar{c}_h(0)\|$ is small enough.

To prove the same with $\bar{c}_{h_1}(0)$, $\bar{c}_h(0)$ switched, we perform the same perturbation argument keeping $c_1 = \bar{c}_h(0)$ and $c_2 = \bar{c}^\mu_{h_1}(0)$. $\qquad\square$

We now perform one more step of perturbation by taking $E > 0$.

Proposition 12.3. *Suppose the slow mechanical system H^s satisfies conditions $[DR1^c] - [DR4^c]$, then there exists $e > 0$ such that the following hold. For each $0 \le E \le e$, there exists a section $S(E) := S(m(E), a(E), \omega(E))$, with $S(E) \to S(0)$ in Hausdorff distance, and the conditions [N1A]–[N3A] are satisfied for $\bar{c}^\mu_{h_1}(E), \bar{c}_h(E)$ at $S(E)$.*

Moreover, the same conditions are satisfied with $\bar{c}_h(E)$ and $\bar{c}^\mu_{h_1}(E)$ switched.

Proof of Proposition 12.3. Since the Aubry sets $\mathcal{A}_{H^s}(\bar{c}_{h_1}(0))$ and $\mathcal{A}_{H^s}(\bar{c}_h(0))$ both have a unique static class, Proposition 7.4 implies that [N2A] and [N3A] are both robust under the perturbation of c_1 and c_2. As a result, it suffices to construct a section $S(E) = S(m(E), \omega(E), a(E))$ such that [N1A] holds, and $S(E) \to S(0)$ as $E \to 0$.

To do this, we choose $\omega(E)$ to be a unit vector orthogonal to $\bar{c}_{h_1}(E) - \bar{c}_h(E)$, and by continuity of the functions $\bar{c}$, $\omega(E) \to \omega(0)$ as $E \to 0$. Since $\alpha_{H^s}(\bar{c}_{h_1}(E)) = \alpha_{H^s}(\bar{c}_h(E)) = E$, [N1A] is satisfied. We then choose $m(E) = m(0)$ and $a(E) = a(0)$. Clearly $S(E) \to S(0)$ in Hausdorff metric. The proposition follows. $\qquad\square$

12.2 VARIATIONAL PROBLEM FOR ORIGINAL COORDINATES

The original Hamiltonian H_ϵ is reduced via coordinate change, and time change, to

$$H^s_\epsilon(\varphi^s, I^s, \tau) = K(I^s) - U(\varphi^s) + \sqrt{\epsilon}\, P(\varphi^s, I^s, \tau),$$

with $\|P\|_{C^2} \le C_1$, see Section 14.2. The system H^s_ϵ is defined on $\mathbb{T}^2 \times \mathbb{R}^2 \times \mathbb{R}$, but is $\sqrt{\epsilon}$ periodic in τ, i.e. it is a periodic Tonelli Hamiltonian introduced in Section 6.1. Denote $\sqrt{\epsilon}\mathbb{T} = \mathbb{R}/(\sqrt{\epsilon}\mathbb{Z})$, then H^s_ϵ projects to $\mathbb{T}^2 \times \mathbb{R}^2 \times \sqrt{\epsilon}\mathbb{T}$.

We now define a variational problem for the perturbed system. First, we will adjust the cohomologies so that they have the same alpha function.

Lemma 12.4. *Fix $e > 0$. There exist $C > 0$, and $\epsilon_0 > 0$ depending only on $K(I)$, $\|U\|_{C^0}$ and e such that for any $\frac{e}{3} \le E \le \frac{2e}{3}$ and $0 \le \epsilon \le \epsilon_0$, there exists*

$0 < E^\epsilon < e$ such that

$$\alpha_{H^s_\epsilon}(\bar{c}_h(E)) = \alpha^\mu_{H^s_\epsilon}(\bar{c}_{h_1}(E^\epsilon)), \quad |E - E^\epsilon| \le C\sqrt{\epsilon}.$$

Proof. We note that $\alpha_{H^s}(c_h(E)) = E = \alpha_{H^s}(c^\mu_{h_1}(E))$. By Lemma 7.9,

$$\left\|\alpha_{H^s_\epsilon}(c_h(E)) - E\right\|, \quad \left\|\alpha_{H^s_\epsilon}(c^\mu_{h_1}(E)) - E\right\| \le C\sqrt{\epsilon}.$$

The Lemma easily follows if we choose ϵ_0 such that $e > 3C\sqrt{\epsilon_0}$. $\square$

We define a section $S^\epsilon(E) = S(m(E), a(E), w^\epsilon(E))$, by keeping $m(E)$, $a(E)$ the same as before, with $w^\epsilon(E)$ to be a unit vector orthogonal to $\bar{c}_h(E) - \bar{c}^\mu_{h_1}(E^\epsilon)$. It is natural to study the extended section $S^\epsilon \times \sqrt{\epsilon}\mathbb{T} \subset \mathbb{T}^2 \times \sqrt{\epsilon}\mathbb{T}$, and condition [N1] becomes

$$\begin{bmatrix} c_1 - c_2 \\ -\alpha_{H^s_\epsilon}(c_1) + \alpha_{H^s_\epsilon}(c_2) \end{bmatrix} \perp S^\epsilon \times \sqrt{\epsilon}\mathbb{T},$$

and the variational problem for [N2] is: For $(x, 0) \in \mathcal{A}_{H^s_\epsilon}(\bar{c}_h(E))$ and $(z, 0) \in \mathcal{A}_{H^s_\epsilon}(\bar{c}^\mu_{h_1}(E^\epsilon))$, consider the minimization

$$\min_{(y,t) \in S^\epsilon \times \sqrt{\epsilon}\mathbb{T}} \left\{ h_{H^s_\epsilon, \bar{c}_h(E)}(x, 0; y, t) + h_{H^s_\epsilon, \bar{c}^\mu_{h_1}(E^\epsilon)}(y, t; z, 0) \right\}. \tag{12.3}$$

We will not, however, study conditions [N1]–[N3] for H^s_ϵ directly. Instead, we transform the cohomology $\bar{c}_h(E)$ and $\bar{c}_{h_1}(E^\epsilon)$, the section $S^\epsilon \times \sqrt{\epsilon}\mathbb{T}$, and the variational problem (12.3) directly to the original system H_ϵ.

Recall that the original Hamiltonian H_ϵ can be brought into a normal form system N_ϵ. N_ϵ is related to H^s_ϵ via the coordinate Φ_L, and an energy reduction, see section 14.2. Denote

$$\Phi^1_L(\varphi, \tau) = B^{-1} \begin{bmatrix} \varphi \\ \tau/\sqrt{\epsilon} \end{bmatrix}, \quad (\Phi^1_L)^{-1}(\theta, t) = \begin{bmatrix} 1 & 0 \\ 0 & \sqrt{\epsilon} \end{bmatrix} B \begin{bmatrix} \theta \\ t \end{bmatrix},$$

this is the angular component of the affine coordinate change Φ_L (see (14.17)). The 3×3 matrix B is defined in (14.10).

Given $\bar{c} \in \mathbb{R}^2$, we define

$$c^\epsilon = p_0 + B_0^T \begin{bmatrix} \sqrt{\epsilon}\bar{c} \\ -\epsilon\alpha_{H^s_\epsilon}(\bar{c}) \end{bmatrix}, \tag{12.4}$$

where B_0 is the first two rows of B, given precisely by k_1^T, k_2^T. According to Proposition 14.9, we have

$$\begin{bmatrix} c^\epsilon - p_0 \\ -\alpha_{H_\epsilon}(c^\epsilon) + H_0(p_0) \end{bmatrix} = B^T \begin{bmatrix} \sqrt{\epsilon}\bar{c} \\ -\epsilon\alpha_{H^s_\epsilon}(\bar{c}) \end{bmatrix} = \Phi^*_L(\bar{c}, \alpha_{H^s}(\bar{c})), \tag{12.5}$$

where Φ^*_L is from (5.8), and coincides with the action component of Φ_L. In particular, (12.4) is the first row of (12.5). Let us denote the cohomologies $c_h^\epsilon(E)$ and $c_{h_1}^\epsilon(E)$ the image of $\bar{c} = \bar{c}_h(E)$ and $\bar{c} = \bar{c}_{h_1}^\epsilon(E)$ under (12.4).

We define a section $\Sigma = \Sigma(\theta_0, a, \Omega, l) \subset \mathbb{T}^2 \times \mathbb{T}$ (for the original Hamiltonian $H_\epsilon : \mathbb{T}^2 \times \mathbb{R}^2 \times \mathbb{T} \to \mathbb{R}$) by

$$\Sigma(\theta_0, a, \Omega, l) = \{(\theta_0 + \lambda\Omega + lt, t) \in \mathbb{T}^2 \times \mathbb{T} : -a < \lambda < a, \ t \in \mathbb{T}\}, \qquad (12.6)$$

where

$$\Omega \in \mathbb{R}^3, \quad l \in \mathbb{Z}^3.$$

The section $S(m, a, \omega) \times \sqrt{\epsilon}\mathbb{T} \subset \mathbb{T}^2 \times \sqrt{\epsilon}\mathbb{T}$ is mapped under Φ_L^1 to $\Sigma(\theta_0, a, \Omega, l)$ with

$$\theta_0 = B^{-1}\begin{bmatrix} m \\ 0 \end{bmatrix} \in \mathbb{T}^3, \quad \Omega = B^{-1}\begin{bmatrix} \omega \\ 0 \end{bmatrix} \in \mathbb{R}^3, \quad l = B^{-1}\begin{bmatrix} 0 \\ 0 \\ 1 \end{bmatrix} \in \mathbb{Z}^3. \qquad (12.7)$$

Note that for $(c_i^\epsilon, \alpha_{H_\epsilon}(c_i^\epsilon)) = \Phi_L^*(\bar{c}_i, \alpha_{H^s}(\bar{c}_i))$, $i = 1, 2$, and using (12.4) and $\Sigma = \Phi_L^1(S \times \sqrt{\epsilon}\mathbb{T})$, we have

$$\begin{bmatrix} \bar{c}_1 - \bar{c}_2 \\ -\alpha_{H^s}(\bar{c}_1) + \alpha_{H^s}(\bar{c}_2) \end{bmatrix} \perp S \times \sqrt{\epsilon}\mathbb{T} \iff \begin{bmatrix} c_1^\epsilon - c_2^\epsilon \\ -\alpha_{H_\epsilon}(c_1^\epsilon) + \alpha_{H_\epsilon}(c_2^\epsilon) \end{bmatrix} \perp \Sigma. \qquad (12.8)$$

We say that $c_1, c_2 \in \mathbb{R}^2$ have non-degenerate connection along a section $\Sigma(\theta, a, \Omega, l)$, if the following conditions hold.

[N1] We have

$$\begin{bmatrix} c_1 - c_2 \\ -\alpha_{H_\epsilon}(c_1) + \alpha_{H_\epsilon}(c_2) \end{bmatrix} \perp \Sigma.$$

[N2] There exists a compact set $K \subset \Sigma$ such that: For each $(x, 0) \in \mathcal{A}_{H_\epsilon}(c_1)$ and $(z, 0) \in \mathcal{A}_{H_\epsilon(c_2)}$, the minimum in

$$\min_{(y,t)\in\Sigma} \{h_{H_\epsilon, c_1}(x, 0; y, t) + h_{H_\epsilon, c_2}(y, t; z, 0)\}$$

is never achieved outside of K.

[N3] Assume that the minimum [N2] is reached at (y_0, t_0), and let $p_1 - c_1$ and $-p_2 + c_2$ be any super-differentials of $h_{c_1}(x, 0; \cdot, t)$ and $h_{c_2}(\cdot, t_0; z, 0)$ respectively. Then

$$(\partial_p H(y_0, p_1, t_0), 1) \cdot \Sigma^\perp, \quad (\partial_p H(y_0, p_2, t_0), 1) \cdot \Sigma^\perp$$

have the same signs, where $\Sigma^\perp$ is a normal vector to Σ.

Proposition 12.5. *Consider $c_h^\epsilon(E), c_{h_1}^\epsilon(E)$ be defined from $\bar{c}_h(E), \bar{c}_{h_1}(E)$ using (12.4). Let E^ϵ be as in Lemma 12.4, and let the section $S^\epsilon(E)$ be as in (12.3), and the section $\Sigma^\epsilon(E)$ be obtained from $S^\epsilon(E) \times \sqrt{\epsilon}\mathbb{T}$ via (12.7).*

Then for each $0 < \epsilon < \epsilon_0$, we have

$$c_h^\epsilon(E), c_{h_1}^\epsilon(E^\epsilon), \Sigma^\epsilon(E)$$

satisfy [N1]–[N3]. Moreover, the same holds with $c_h^\epsilon(E)$ and $c_{h_1}^\epsilon(E^\epsilon)$ switched.

The following statement is a direct application of Theorem 12.8, which holds for general Tonelli Hamiltonians.

Proposition 12.6. *Assume that the conclusions of Proposition 12.5 hold. In addition, assume that both $\mathcal{A}_{H_\epsilon}(c_h^\epsilon(E))$ and $\mathcal{A}_{H_\epsilon}(c_{h_1}^\epsilon(E^\epsilon))$ admit a unique static class. Then*

$$c_h(E) \dashv\vdash c_{h_1}^{\epsilon,\mu}(E^\epsilon).$$

Proof of Theorem 5.9. By Proposition 12.5, for the system H_ϵ, which is a perturbation of H_ϵ, all conditions of Proposition 12.6 are satisfied, except the condition of uniqueness of static classes. For this we consider again the residual condition that all rational minimal periodic orbits are hyperbolic, and have transversal homoclinic and heteroclinic intersections. Under this condition all cohomologies $c_h(E)$ and $c_{h_1}^{\epsilon,\mu}(E)$ have a unique static class (using the fact that they are of Aubry-Mather type). □

We prove Proposition 12.5 in Section 12.3 and Proposition 12.6 in Section 12.4.

12.3   SCALING LIMIT OF THE BARRIER FUNCTION

In this section we prove Proposition 12.5. Using the choice of the cohomology classes and the sections, together with (12.8), we get [N1] is satisfied for $c_h^\epsilon(E)$, $c_{h_1}^\epsilon(E^\epsilon)$, and $\Sigma^\epsilon(E)$. It suffices to prove [N2] and [N3]. We will show that the variational problem in [N2] is a scaling limit of the variational problem (12.3).

Proposition 12.7. *The family of functions $h_{H_\epsilon, c_h(E)}(\chi^{E,\epsilon}, 0; \cdot, t)/\sqrt{\epsilon}$ is uniformly semi-concave, and*

$$\lim_{\epsilon \to 0+} \sup_{(\theta,t) \in \mathbb{T}^2 \times \mathbb{T}} \left| h_{H_\epsilon, c_h^\epsilon(E)}(\chi^{E,\epsilon}, 0; \theta, t)/\sqrt{\epsilon} - h_{H^s, \bar{c}_h(E)}(x^E, B_0 \begin{bmatrix} \theta \\ t \end{bmatrix}) \right| = 0$$

uniformly over

$$(\chi^{E,\epsilon}, 0) \in \mathcal{A}_{H_\epsilon}(c_h(E)), \quad (x^E, 0) \in \mathcal{A}_{H_\epsilon}(\bar{c}_h(E)).$$

The same conclusions apply to the barrier function $h_{H_\epsilon, c_{h_1}^\epsilon(E)}(\cdot, \cdot; \chi^{E,\epsilon}, 0)/\sqrt{\epsilon}$.

Proof. The uniform semi-concavity of barrier functions follows from Proposition 7.6. Moreover, according to Proposition 14.9, item 2, the families of functions

$$h_{H_\epsilon, c_h^\epsilon(E)}(\chi^{E,\epsilon}, 0; \theta, t)/\sqrt{\epsilon}, \qquad h_{H_\epsilon^s, \bar{c}_h(E)}\left(\Phi_{LS}^1(\chi^{E,\epsilon}, 0); \Phi_L^1(\theta, t)\right)$$

share the same limit points as $\epsilon \to 0$. Moreover, Proposition 14.9, item 3 implies $\Phi_L^1(\chi^{E,\epsilon}, 0) \in \mathcal{A}_{H_\epsilon^s}(\bar{c}_h(E))$.

The functions H_ϵ^s form a uniform family of periodic Hamiltonians, and $H_\epsilon^s \to H^s$ in $C^2(\mathbb{T}^2 \times \mathbb{R}^2 \times \mathbb{R})$. By construction, $\mathcal{A}_{H^s}(\bar{c}_h(E))$ has unique static class; therefore Proposition 7.4 applies, and for any $(x_E, 0) \in \mathcal{A}_{H^s}(\bar{c}_h(E))$,

$$h_{H_\epsilon^s, \bar{c}_h(E)}\left(\Phi_{LS}^1(\chi^{E,\epsilon}, 0); \varphi, \tau)\right) \to h_{H^s, \bar{c}_h(E)}\left(x^E, 0; \varphi, \tau\right)$$

uniformly. Finally, noticing

$$\Phi_L^1(\theta, t) = \begin{bmatrix} \mathrm{Id} & 0 \\ 0 & \sqrt{\epsilon} \end{bmatrix} B \begin{bmatrix} \theta \\ t \end{bmatrix} = \begin{bmatrix} \mathrm{Id} & 0 \\ 0 & \sqrt{\epsilon} \end{bmatrix} \begin{bmatrix} B_0 \\ k_3^T \end{bmatrix} \begin{bmatrix} \theta \\ t \end{bmatrix} \to \begin{bmatrix} B_0 \begin{bmatrix} \theta \\ t \end{bmatrix} \\ 0 \end{bmatrix} \tag{12.9}$$

as $\epsilon \to 0$, we obtain the desired limit. The case for $h_{H_\epsilon, c_{h_1}^\epsilon(E)}(\cdot, \cdot; \chi^{E,\epsilon}, 0)/\sqrt{\epsilon}$ is symmetric and we omit it. □

Proof of Proposition 12.5. Condition [N1] is satisfied by the choice of section. Observe that

$$B_0\Sigma^\epsilon(E) = \begin{bmatrix} \mathrm{Id} & 0 \\ 0 & 0 \end{bmatrix} B\Sigma^\epsilon(E) = \begin{bmatrix} \mathrm{Id} & 0 \\ 0 & 0 \end{bmatrix} BB^{-1} \begin{bmatrix} \mathrm{Id} & 0 \\ 0 & \sqrt{\epsilon}^{-1} \end{bmatrix} (S^\epsilon(E) \times \sqrt{\epsilon}\mathbb{T})$$

$$= \begin{bmatrix} \mathrm{Id} & 0 \\ 0 & 0 \end{bmatrix} (S^\epsilon(E) \times \sqrt{\epsilon}\mathbb{T}) = S^\epsilon(E).$$

Since $S^\epsilon(E) \to S(E)$ as $E \to 0$, combing with Proposition 12.3 we get [N2] still holds for $0 \le \epsilon < \epsilon_0$, with ϵ_0 depending only on H^s. [N3] follows from the same limiting argument and we omit it. □

12.4 THE JUMP MECHANISM

In this section we show that the [N1]–[N3] condition combined with uniqueness of static class imply $c_1 \vdash c_2$, which is Proposition 12.6. The discussions here apply to general Tonelli Hamiltonians. Denote

$$\tilde{A}_{H,c}(x, s, y, t) = A_{H,c}(x, s, y, t) + \alpha_H(c)(t - s).$$

The subscript H may be omitted.

Theorem 12.8. *Assume that c_1, c_2 and Σ satisfy the conditions [N1]–[N3], and in addition,*

$$\mathcal{A}_H(c_1), \quad \mathcal{A}_H(c_2)$$

both have unique static class, then the following hold.

1. *(Interior Minimum) There exist $N < N', M < M' \in \mathbb{N}$ and a relatively compact set $K' \subset \Sigma$, such that for any semi-concave function u on $\mathbb{T}^2$, the minimum in*

$$v(z) := \min\{u(x) + \tilde{A}_{c_1}(x, 0; y, t+n) + \tilde{A}_{c_2}(y, t+n; z, n+m)\}, \quad (12.10)$$

where the minimum is taken in

$$x \in \mathbb{T}^2, \ (y, t) \in \Sigma, \ N \le n \le N', \ M \le m \le M',$$

is never achieved for $(y, t) \notin K'$.
2. *(No Corner) Assume the minimum in (12.10) is achieved at $(y, t) = (y_0, t_0)$, $(n, m) = (n_0, m_0)$, and the minimizing curves are $\gamma_1 : [0, t_0 + n_0] \to \mathbb{T}^2$ and $\gamma_2 : [t_0 + n_0, t_0 + n_0 + m_0] \to \mathbb{T}^2$. Then γ_1 and γ_2 connect to an orbit of the Euler-Lagrange equation, i.e.*

$$\dot{\gamma}_1(t_0 + n_0) = \dot{\gamma}_2(t_0 + n_0).$$

3. *(Connecting Orbits) The function v is semi-concave, and its associated pseudograph satisfies*

$$\overline{\mathcal{G}_{c_2,v}} \subset \bigcup_{0 \le t \le N'+M'} \phi^t \mathcal{G}_{c_1,u}.$$

As a consequence,

$$c_1 \vdash c_2.$$

Remark 12.9. We only need item 3 of the theorem for our purpose, the first two items are stated to illuminate the idea. Item 1 can be seen as a finite time version of the variational problem in [N2]. It holds for sufficiently large M, N, due to uniform convergence of the Lax-Oleinik semi-group. Item 2 implies the minimizers concatenate to a real orbit of the system, which is crucial in proving item 3.

Lemma 12.10. *1. Let u be a continuous function on $\mathbb{T}^2$. The limit*

$$\lim_{N \to \infty} \lim_{N' \to \infty} \min_{x \in \mathbb{T}^2, N \le n \le N'} \{u(x) + \tilde{A}_c(x, 0; y, t+n)\}$$

$$= \min_{x \in \mathbb{T}^2} \{u(x) + h_c(x, 0; y, t)\}$$

is uniform in u and (y, t).

2. *The limit*

$$\lim_{N\to\infty} \lim_{N'\to\infty} \min_{N\leq n\leq N'} \tilde{A}_c(y,t;z,n) = h_c(y,t;z,0)$$

is uniform in y, t, z.

Proof. The proof of the first item is similar to the proof of Proposition 6.3 of [11] with some auxiliary facts proven in Appendix A there. The proof of the second item is similar to that of Proposition 6.1 from [11].

In both cases the action function, defined in (2.4) and (6.1) of [11], is restricted an to have integer time increment. For non-integer time increments the same argument applies. □

Using the representation formula (Proposition 6.16) and Lemma 7.5, we have the following characterization of the barrier functions.

Lemma 12.11. *Assume that* $\mathcal{A}(c)$ *has only one static class. For each point* $(y,t) \in \mathbb{T}^2 \times \mathbb{T}$ *and each* $z \in \mathbb{T}^2$

1. *There exist* $x_0 \in \mathbb{T}^2$ *and* $x_1 \in \mathcal{A}(c)$ *such that*

$$\min_{x\in\mathbb{T}^2} \{u(x) + h_c(x,0;y,t)\} = u(x_0) + h_c(x_0,0;x_1,0) + h_c(x_1,0;y,t).$$

2. *There exists* $z_1 \in \mathcal{A}(c)$ *such that*

$$h_c(y,t;z,0) = h_c(y,t;z_1,0) + h_c(z_1,0;z,0).$$

Proof of Theorem 12.8. According to Lemma 12.10, (12.10) converges uniformly as $N, M \to \infty$ to

$$\min_{x,y,t} \{u(x) + h_{c_1}(x,0;y,t) + h_{c_2}(y,t;z,0)\},$$

which is equal to

$$\min_{(y,t)} \{u(x_0) + h_{c_1}(x_0,0;x_1,0) + h_{c_1}(x_1,0;y,t) + h_{c_2}(y,t;z_1,0) + h_{c_2}(z_1,0;z,0)\}$$

$$= \min_{(y,t)} \{const + h_{c_1}(x_1,0;y,t) + h_{c_2}(y,t;z_1,0) + h_{c_2}(z_1,0;z,0)\}.$$

by Lemma 12.11. Since the above variational problem has a interior minimum due to condition N2, by uniform convergence, the finite-time variational problem (12.10) also has an interior minimum for sufficiently large N, M.

We now prove the second conclusion. Let γ_1 and γ_2 be the minimizers for $\tilde{A}_{c_1}(x_0,0;y_0,t_0+n_0)$ and $\tilde{A}_{c_2}(y_0,t_0+n_0;z,n_0+m_0)$, and let p_1 and p_2 be the associated momentum. We will show that

$$p_1(t_0+n_0) = p_2(t_0+n_0),$$

which implies $\dot{\gamma}_1(t_0 + n_0) = \dot{\gamma}_2(t_0 + n_0)$. To abbreviate notations, we write $p_1^0 = p_1(t_0 + n_0)$ and $p_2^0 = p_2(t_0 + n_0)$ for the rest of the proof.

Note that

$$u_1(x_0) + \tilde{A}_{c_1}(x_0, 0; y_0, t_0 + n_0) = \min_{x \in \mathbb{T}^2}\{u_1(x) + \tilde{A}_{c_1}(x, 0; y_0, t_0 + n_0)\}.$$

By semi-concavity, the function $u_1(x) + \tilde{A}_{c_1}(x, 0; y_0, t_0 + n_0)$ is differentiable at x_0 and the derivative vanishes. By Proposition 6.2 part 3,

$$d_x u(x_0) = p_1(0) - c_1. \tag{12.11}$$

By a similar reasoning, we have

$$\tilde{A}_{c_1}(x_0, 0; y_0, t_0 + n_0) + \tilde{A}_{c_2}(y_0, t_0 + n_0; z, n_0 + m_0)$$
$$= \min_{(y,t) \in \overline{\Sigma}}\{\tilde{A}_{c_1}(x_0, 0; y, t + n_0) + \tilde{A}_{c_2}(y, t + n_0; z, n_0 + m_0)\}. \tag{12.12}$$

By Proposition 6.2, we know

$$(p_1^0 - c_1, \alpha_H(c_1) - H(y_0, t_0, p_0^1)), \quad (-p_2^0 + c_1, -\alpha_H(c_2) + H(y_0, t_0, p_0^2))$$

are super-differentials of $\tilde{A}_{c_1}(x_0, 0; y_0, t_0 + n_0)$ and $\tilde{A}_{c_2}(y_0, t_0 + n_0; z, n_0 + m_0)$ at $(y_0, t_0 + n_0)$, since the minimum in (12.12) is obtained inside of Σ, we have

$$(p_1^0 - p_2^0, -H(y_0, t_0, p_1^0 + H(y_0, t_0, p_2^0)) + (-c_1 + c_2, \alpha_H(c_1) - \alpha_H(c_2))$$

is orthogonal to Σ. Since the second term is orthogonal to Σ by [N1], we obtain

$$(p_1^0 - p_2^0, -H(y_0, t_0, p_1^0) + H(y_0, t_0, p_2^0)) \in \mathbb{R}\Sigma^\perp.$$

We proceed to prove $p_1^0 = p_2^0$.

Write $\Sigma^\perp = (v, w) \in \mathbb{R}^2 \times \mathbb{R}$, then there exists λ_0 such that

$$p_1^0 - p_2^0 = \lambda_0 v, \quad -H(y_0, t_0, p_1^0) + H(y_0, t_0, p_2^0) = \lambda_0 w.$$

Define

$$f(\lambda) = H(y_0, t_0, p_1^0 + \lambda v) - H(y_0, t_0, p_1) + \lambda w;$$

then $f(\lambda_0) = 0$. Since $f(\lambda)$ is strictly convex on $\mathbb{R}$, there are at most two solutions to $f(\lambda) = 0$, one of them is $\lambda = 0$. Suppose $\lambda_0 \neq 0$, by strict convexity again, $f'(0)$ and $f'(\lambda_0)$ must have different signs. Since

$$f'(0) = (\partial_p H(y_0, t_0, p_1^0), 1) \cdot (v, w), \quad f'(\lambda_0) = (\partial_p H(y_0, t_0, p_2^0), 1) \cdot (v, w),$$

have the same signs from [N3], we get a contradiction. As a result, 0 is the only solution to $f(\lambda) = 0$, indicating $p_1^0 = p_2^0$. Moreover, this implies $p_1^0 = p_2^0$ is uniquely defined and the functions $\tilde{A}_{c_1}(x_0, 0; \cdot, \cdot)$, $\tilde{A}_{c_2}(\cdot, \cdot; z_0, n_0 + m_0)$ are differentiable at $(y_0, t_0 + n_0)$.

As a consequence, (γ_1, p_1) and (γ_2, p_2) connect as a solution of the Hamiltonian flow. Using (12.11), we have

$$\phi^{n_0+m_0}(x_0, du_x(x_0) + c_1) = (z, p_2(n_0 + m_0)).$$

Note that $p_2(n_0 + m_0) - c_2$ is a super-differential to v at z. If v is differentiable at z, then $p_2 = dv(z) + c_2$. This implies

$$\overline{\mathcal{G}_{c_2,v}} \subset \bigcup_{0 \le t \le N'+M'} \phi^t \mathcal{G}_{c_1,u}$$

and the forcing relation.   □

Part IV

Supplementary topics

Chapter Thirteen

Generic properties of mechanical systems on the two-torus

This chapter is devoted to proving generic properties of the minimizing orbits of the slow mechanical system. In the first three sections, we prove Theorem 4.5 concerning non-critical but bounded energy. In Section 13.4 we prove Proposition 4.6 concerning the very high energy, and in Section 13.5 we prove Proposition 4.7 concerning the critical energy.

The proof of Theorem 4.5 consists of three steps.

1. In Section 13.1, we prove a Kupka-Smale-like theorem about non-degeneracy of periodic orbits. For a fixed energy surface, generically, all periodic orbits are non-degenerate. This fails for an interval of energies. We show that while degenerate periodic orbits exist, there are only finitely many of them. Moreover, there could be only a particular type of bifurcation for any family of periodic orbits crossing a degeneracy.
2. In Section 13.2, we show that a non-degenerate locally minimal orbit is always *hyperbolic*. Using step 1, we show that for each energy, the globally minimal orbits is chosen from a finite family of *hyperbolic* locally minimal orbits.
3. In Section 13.3, we finish the proof by proving the finite local families obtained from step 2 are "in general position," and therefore there are at most two global minimizers for each energy.

13.1 GENERIC PROPERTIES OF PERIODIC ORBITS

We simplify notations and drop the superscript "s" from the notation of the slow mechanical system. Moreover, we treat U as a parameter, and write

$$H^U(\varphi, I) = K(I) - U(\varphi), \quad \varphi \in \mathbb{T}^2, \ I \in \mathbb{R}^2, \ U \in C^r(\mathbb{T}^2). \tag{13.1}$$

We shall use U as *an infinite-dimensional parameter*. Denote by $\mathcal{G}^r = C^r(\mathbb{T}^2)$ the space of potentials, x denotes (φ, I), W denotes $\mathbb{T}^2 \times \mathbb{R}^2$, and either ϕ_t^U or $\Phi(\cdot, t, U)$ denotes the flow of (13.1). We will use $\chi^U(x) = (\partial K, \partial U)(x)$ to denote the Hamiltonian vector field of H^U and use S_E to denote the energy surface $\{H^U = E\}$. We may drop the superscript U when there is no confusion.

By the invariance of the energy surface, the differential map $D_x \phi_t^U$ defines a map

$$D_x \phi_t^U(x) : T_x S_{H(x)} \to T_{\phi_t^U(x)} S_{H(x)}.$$

Since the vector field $\chi(x)$ is invariant under the flow, $D_x \phi$ induces a factor map

$$\bar{D}_x \phi_t^U(x) : T_x S_{H(x)} / \mathbb{R}\chi(x) \to T_{\phi_t^U(x)} S_{H(x)} / \mathbb{R}\chi(\phi_t(x)).$$

Given $U_0 \in \mathcal{G}^r$, $x_0 \in W$ and $t_0 \in \mathbb{R}$, let

$$\mathcal{V} = V(x_0) \times (t_0 - a, t_0 + a) \times V(U_0)$$

be a neighborhood of (x_0, t_0, U_0), $V(x_0, t_0)$ of (x_0, t_0), and $V(\phi_{t_0}^{U_0}(x_0))$ a neighborhood of $\phi_{t_0}^{U_0}(x_0)$, such that

$$\phi_t^U(x) \in V(\phi_{t_0}^{U_0}(x_0)), \quad (x, t, U) \in \mathcal{V}.$$

By fixing the local coordinates on $V(x_0)$ and $V(\phi_{t_0}^U(x_x))$, we define

$$\tilde{D}_x \Phi : \mathcal{V} \to Sp(2),$$

where $\tilde{D}_x \Phi(x, t, U)$ is the 2×2 symplectic matrix associated to $\bar{D}_x \varphi_t^U(x)$ under the local coordinates. The definition depends on the choice of coordinates.

Let $\{\phi_t^{U_0}(x_0)\}$ be a periodic orbit with minimal period t_0. The periodic orbit is *non-degenerate* if and only if 1 is *not* an eigenvalue of $\tilde{D}_x \Phi(x_0, t_0, U_0)$. Furthermore, a degenerate periodic orbit necessarily falls into one of two types:

1. A degenerate periodic orbit (x_0, t_0, U_0) is of *type I* if $\tilde{D}_x \Phi(x_0, t_0, U_0) = Id$, the identity matrix;
2. It is of *type II* if $\tilde{D}_x \Phi(x_0, t_0, U_0)$ is conjugate to the matrix $[1, \mu; 0, 1]$ for $\mu \neq 0$.

Denote

$$N = \left\{ \begin{bmatrix} 1 & \mu \\ 0 & 1 \end{bmatrix} : \mu \in \mathbb{R} \setminus \{0\} \right\}, \quad \mathcal{O}(N) = \{BAB^{-1} : A \in N, B \in Sp(2)\}.$$

Then (x_0, t_0, U_0) is of type II if and only if $\tilde{D}_x \Phi(x_0, t_0, U_0) \in \mathcal{O}(N)$.

Lemma 13.1. *The set $\mathcal{O}(N)$ is a 2-dimensional submanifold of $Sp(2)$.*

Proof. Any matrix in $\mathcal{O}(N)$ can be expressed by

$$\begin{bmatrix} a & b \\ c & d \end{bmatrix} \begin{bmatrix} 1 & \mu \\ 0 & 1 \end{bmatrix} \begin{bmatrix} d & -b \\ -c & a \end{bmatrix} = \begin{bmatrix} 1 - ac\mu & a^2\mu \\ -c^2\mu & 1 - ac\mu \end{bmatrix},$$

where $ad - bc = 1$ and $\mu \neq 0$. Writing $\alpha = a^2\mu$ and $\beta = ac\mu$, we can express any matrix in $\mathcal{O}(N)$ by

$$\begin{bmatrix} 1 - \beta & \alpha \\ 1 - \beta^2/\alpha & 1 - \beta \end{bmatrix}. \tag{13.2}$$

$\square$

The standard Kupka-Smale theorem (see [69, 71]) *no longer holds for an interval of energies*. Generically, periodic orbits appear in one-parameter families and may contain degenerate ones. However, while degenerate periodic orbits may appear, generically, a family of periodic orbits crosses the degeneracy "transversally". This is made precise in the following theorem.

Theorem 13.2. *Suppose $r \geq 3$. There exists a residual subset $\mathcal{G}'$ of $\mathcal{G}^r$, such that for all $U \in \mathcal{G}'$, the following hold:*

1. *The set of periodic orbits for ϕ_t^U form a submanifold of dimension 2. Since a periodic orbit itself is a 1-dimensional manifold, distinct periodic orbits form one-parameter families.*
2. *There are no degenerate periodic orbits of type I.*
3. *The set of periodic orbits of type II forms a 1-dimensional manifold. Factoring out the flow direction, the set of type II degenerate orbits are isolated.*
4. *For $U_0 \in \mathcal{G}^r$, let $\Lambda^{U_0} \subset W \times \mathbb{R}^+$ denote the set of periodic orbits for ϕ_t^U, and $\Lambda_N^{U_0} \subset \Lambda^{U_0}$ denote the set of type II degenerate ones. Then for any $(x_0, t_0) \in \Lambda_N^{U_0}$, let $V(x_0, t_0)$ be a neighborhood of (x_0, t_0). Then*

$$\tilde{D}_x \Phi|_{U=U_0} : \Lambda^{U_0} \cap V(x_0, t_0) \to Sp(2)$$

is transversal to $\mathcal{O}(N) \subset Sp(2)$.

Remark 13.3. Statement 4 of the theorem can be interpreted in the following way. Let $A(\lambda)$ be the differential of the Poincaré return map on associated to a family of periodic orbits. Then if $A(\lambda_0) \in \mathcal{O}(N)$, then the tangent vector $A'(\lambda_0)$ is transversal to $\mathcal{O}(N)$.

We can improve the set $\mathcal{G}'$ to an open and dense set, if there is a lower and upper bound on the minimal period.

Corollary 13.4. *1. Given $0 < T_0 < T_1$, there exists an open and dense subset $\mathcal{G}' \subset \mathcal{G}^r$, such that the set of periodic orbits with minimal period in $[T_0, T_1]$ satisfies the conclusions of Theorem 13.2.*
2. For any $U_0 \in \mathcal{G}'$, there are at most finitely many type II degenerate periodic orbits. Moreover, there exists a neighborhood $V(U_0)$ of U_0, such that the set of type II degenerate periodic orbits depends smoothly on U. (This means the number of such periodic orbits is constant on $V(U_0)$, and each periodic orbit depends smoothly on U.)

We define

$$F : W \times \mathbb{R}^+ \times \mathcal{G}^r \to W \times W, \tag{13.3}$$

$$F(x, t, U) = (x, \Phi(x, t, U)).$$

F is a C^{r-1}-map of Banach manifolds. Define the diagonal set by $\Delta = \{(x, x)\} \subset W \times W$. Then $\{\phi_t^{U_0}(x_0)\}$ is a period orbit of period t_0 if and only if $(x_0, t_0, U_0) \in F^{-1}\Delta$.

Proposition 13.5. *Assume that $(x_0, t_0, U_0) \in F^{-1}\Delta$ or, equivalently, x_0 is periodic orbit of period t_0 for H^U and that t_0 is the minimal period, then there exists a neighborhood $\mathcal{V}$ of (x_0, t_0, U_0) such that the map*

$$d\pi_\Delta^\perp D_{(x,t,U)} F : T_{(x,t,U)}(W \times \mathbb{R}^+ \times \mathcal{G}^r) \to T_{F(x,t,U)}(W \times W)/T\Delta$$

has co-rank 1 for each $(x, t, U) \in \mathcal{V}$, where $d\pi_\Delta^\perp T(W \times W) \to T(W \times W)/T\Delta$ is the standard projection along $T\Delta$.

Remark 13.6. If the aforementioned map has full rank, then the map is called transversal to Δ at (x_0, t_0, U_0). However, the transversality condition fails for our map.

Given $\delta U \in \mathcal{G}^r$, the directional derivative $D_U \Phi \cdot \delta U$ is defined as follows $\frac{\partial}{\partial \epsilon}|_{\epsilon=0} \Phi(x, t, U + \epsilon \delta U)$. The differential $D_U \Phi$ then defines a map from $T\mathcal{G}^r$ to $T_{\Phi(x,t,U)} W$. The following hold for this differential map:

Lemma 13.7. *[69] Assume that there exists $\epsilon > 0$ and $t_0 \in (0, \tau)$ such that the orbit of x has no self-intersection for the time interval $(t_0 - \epsilon, t_0 + \epsilon)$; then the map*

$$D_U \Phi(x, \tau, U) : \mathcal{G}^r \to T_{\Phi(x,\tau,U)} W$$

generates a subspace orthogonal to the gradient $\nabla H^U(\Phi(x, \tau, U))$ and the Hamiltonian vector field $\chi^U(\Phi(x, \tau, U))$ of H^U.

Proof. We refer to Lemmas 16 and 17 of [69]. We note that while the proof was written for a periodic orbit of minimal period τ, the proof holds for non-self-intersecting orbit. □

Proof of Proposition 13.5. We note that if $\{\phi_t^U(x)\}$ is a periodic orbit of minimal period τ, then the orbit $\{\phi_t^{U'}(x')\}$ satisfies the assumptions of Lemma 13.7. It follows that the matrix

$$d\pi_\Delta^\perp \circ D_U F = \begin{bmatrix} D_x \Phi - I & D_t \Phi & D_U \Phi \end{bmatrix}$$

has co-rank 1, since the last two components generates the subspace

$$\text{Image} (D_U \Phi) + \mathbb{R}\chi^U,$$

which is a subspace complementary to ∇H^U. □

Proposition 13.5 allows us to apply the constant rank theorem in Banach spaces.

Proposition 13.8. *The set $F^{-1}\Delta$ as a subset of a Banach space is a subman-ifold of codimension 3. If $r \geq 3$, then for generic $U \in \mathcal{G}^r$, $F^{-1}\Delta \cap \pi_U^{-1}\{U\}$ is a 2-dimensional manifold.*

Proof. We note that the kernel and cokernel of the map $d\pi \circ D_U F$ has finite codimension; hence the constant rank theorem (see Theorem 2.5.15 of [1]) ap-plies. As a consequence, we may assume that locally, $\Delta = \Delta_1 \times (-a, a)$ and that the map $\pi_1 \circ F$ has full rank. Since the dimension of Δ_1 is 3, $F^{-1}\Delta$ is a submanifold of codimension $2n - 1$. The mapping $\pi_U : F^{-1}\Delta \to \mathcal{G}^r$ is Fredholm with index 2. Sard's theorem applies if $r \geq 3$, and at every regular value of π_U, the set $F^{-1}(\Delta) \cap \pi^{-1}\{U\}$ is a submanifold of dimension 2.   □

Denote $\Lambda = F^{-1}\Delta$. On a neighborhood $\mathcal{V}$ of each $(x_0, t_0, U_0) \in \Lambda$, we define the map

$$\tilde{D}_x\Phi : \Lambda \cap \mathcal{V} \to Sp(2), \quad \tilde{D}_x\Phi(x, t, U) = \tilde{D}\phi_t^U(x). \tag{13.4}$$

First we refer to the following lemma by Oliveira:

Lemma 13.9 (Theorem 18 of [69]). *For each $(x_0, t_0, U_0) \in \Lambda$ such that t_0 is the minimal period, let $\tilde{\mathcal{G}}$ be the set of tangent vectors in $T_{(x_0, t_0, U_0)}\Lambda$ of the form $(0, 0, V)$. Then the map*

$$D_U \tilde{D}_x\Phi : \tilde{\mathcal{G}} \to T_{\tilde{D}_x\Phi(x_0, t_0, U_0)} Sp(2)$$

has full rank.

Corollary 13.10. *The map* (13.4) *is transversal to the submanifold $\{Id\}$ and $\mathcal{O}(N)$ of $Sp(2)$.*

Denote

$$\Lambda_{Id} = \Lambda \cap \tilde{D}_x\Phi)^{-1}(\{Id\}) \text{ and } \Lambda_N = \Lambda \cap \tilde{D}_x\Phi)^{-1}(\mathcal{O}(N)).$$

Note that the expression is well defined because both preimages are defined independent of local coordinate changes.

Proof of Theorem 13.2. The first statement of the theorem follows from Propo-sition 13.8.

As the subset $\{Id\}$ has codimension 3 in $Sp(2)$, Λ_{Id} has codimension 3 in Λ, and hence has codimension 6 in $W \times \mathbb{R}^+ \times \mathcal{G}^r$. By Sard's lemma, for a generic $U \in \mathcal{G}^r$, the set $\Lambda_{Id} \cap \pi_U^{-1}$ is empty. This proves the second statement of the theorem.

Since the set $\mathcal{O}(N)$ has codimension 1, Λ_N has codimension 1 in Λ, and hence has codimension 4 in $W \times \mathbb{R}^+ \times \mathcal{G}^r$. As a consequence, generically, the set $\Lambda_N^U = \Lambda_N \cap \pi_U^{-1}$ has dimension 1. This proves the third statement.

Fix $U_0 \in \mathcal{G}'$, the set $\Lambda^{U_0} = \Lambda \cap \pi_U^{-1}(U_0)$ has dimension 2, while $\Lambda_N^{U_0} \subset \Lambda^{U_0}$ has dimension 1. It follows that at any $(x_0, t_0) \in \Lambda_N^{U_0}$, there exists a tangent vector

$$(\delta x, \delta t) \in T_{(x_0,t_0)} \Lambda^{U_0} \setminus T_{(x_0,t_0)} \Lambda_N^{U_0}$$

such that

$$(\delta x, \delta t, 0) \in T_{(x_0,t_0,U_0)} \Lambda \setminus T_{(x_0,t_0,U_0)} \Lambda_N.$$

It follows that

$$D_{(x,t)} \widetilde{D}_x \Phi|_{U=U_0}(x_0, t_0) = D_{(x,t,U)} \widetilde{D}_x \Phi(x_0, t_0, U_0) \cdot (\delta x, \delta t, 0)$$

is not tangent to $\mathcal{O}(N)$. Since $\mathcal{O}(N)$ has codimension 1, this implies that the map $\widetilde{D}_x \Phi|_{U=U_0}$ is transversal to $\mathcal{O}(N)$. This proves the fourth statement. □

Proof of Corollary 13.4. If a potential $U \in \mathcal{G}'$, then by Theorem 13.2 conditions 1–4 are satisfied. In particular, all periodic orbits are either non-degenerate or degenerate satisfying conditions 3 and 4. Non-degenerate periodic orbits of period bounded both from zero and infinity form a compact set. Therefore, they stay non-degenerate for all potential C^r-close to U. By condition 3 degenerate periodic orbits are isolated. This implies that there are finitely many of them. Condition 4 is a transversality condition, which is C^r open for each degenerate orbit. □

Fix $U \in \mathcal{G}'$ as in Corollary 13.4. It follows that periodic orbits of ϕ_t^U form one-parameter families. We now discuss the generic bifurcation of such a family at a degenerate periodic orbit.

Proposition 13.11. *Let (x_λ, t_λ) be a family of periodic orbits such that (x_0, λ_0) is degenerate. On one side of $\lambda = \lambda_0$, say $\lambda > \lambda_0$, the matrix $\widetilde{D}_x \phi_{t_\lambda}^U(x_\lambda)$ has a pair of distinct real eigenvalues; if $\lambda < \lambda_0$, it has a pair of complex eigenvalues.*

Proof. Write $A(\lambda) = \widetilde{D}_x \phi_{t_\lambda}^U(x_\lambda)$ for short. By choosing a proper local coordinates we may assume that $A(\lambda_0) = [1, \mu; 0, 1]$. The tangent space to $Sp(2)$ at $[1, \mu; 0, 1]$ is given by the set of traceless matrices $[a, b; c, -a]$. Using (13.2), we have a basis of the tangent space to $\mathcal{O}(N)$ at $[1, \mu; 0, 1]$ given by

$$\begin{bmatrix} 0 & 1 \\ -\beta^2/\alpha^2 & 0 \end{bmatrix} \quad \text{and} \quad \begin{bmatrix} -1 & 0 \\ -2\beta/\alpha & -1 \end{bmatrix}.$$

An orthogonal matrix to this space, using the inner product $tr(A^T B)$, is given by $[0, 0; 1, 0]$. As a consequence, a matrix $[a, b; c, -a]$ is transversal to $\mathcal{O}(N)$ if and only if $c \neq 0$.

The eigenvalues of the matrix

$$\begin{bmatrix} 1 + ah & \mu + bh \\ ch & 1 - ah \end{bmatrix}$$

are given by $\lambda = 1 \pm \sqrt{a^2 h^2 - bch^2 - \mu ch}$. Using $\mu \neq 0$ and $c \neq 0$ we obtain that $a^2h^2 - bch^2 - \mu ch$ changes sign at $h = 0$. This proves our proposition. $\quad\square$

13.2   GENERIC PROPERTIES OF MINIMAL ORBITS

In this section, we analyze properties of families of minimizing orbits. It has been known since Poincaré that a non-degenerate minimizing geodesic on a surface is hyperbolic. (We give a proof in Proposition 13.12.) However, we have shown in the previous section that degenerate ones do exist. The main idea to avoid degenerate minimizing orbits is the following: When one extends a family of hyperbolic minimal orbits, the family must terminate at a degenerate orbit of type II. We then can "slide" different families against each other so that the degenerate orbit is never the shortest.

Let d_E denote the metric derived from the Riemannian metric $g_E(\varphi, v) = 2(E+U(\varphi))K^{-1}(v)$. We define the arclength of any continuous curve $\gamma : [t, s] \to \mathbb{T}^2$ by

$$l_E(\gamma) = \sup \sum_{i=0}^{N-1} d_E(\gamma(t_i), \gamma(t_{i+1})),$$

where the supremum is taken over all partitions $\{[t_i, t_{i+1}]\}_{i=0}^{N-1}$ of $[t, s]$. A curve γ is called rectifiable if $l_E(\gamma)$ is finite.

A curve $\gamma : [a, b] \to \mathbb{T}^2$ is called piecewise regular, if it is piecewise C^1 and $\dot\gamma(t) \neq 0$ for all $t \in [a, b]$. A piecewise regular curve is always rectifiable.

We write

$$l_E(h) = \inf_{\xi \in \mathcal{C}_h^E} l_E(\xi),$$

where $\mathcal{C}_h^E$ denote the set of all piecewise regular closed curves with homology $h \in H_1(\mathbb{T}^2, \mathbb{Z})$. A curve realizing the infimum is the shortest geodesic curve in the homology h, which we will also refer to as a global (g_E, h)-minimizer.

It is well known that for any $E > -\min_\varphi U(\varphi)$, a global g_E-minimizer is a closed g_E-geodesic. Hence, it corresponds to a periodic orbit of the Hamiltonian flow.

It will also be convenient to consider the local minimizers. Letting $V \subset \mathbb{T}^2$ be an open set, a closed continuous curve $\gamma : [a, b] \to V$ is a local minimizer if

$$l_E(\gamma) = \inf_{\xi \in \mathcal{C}_h^E, \xi \subset \bar V} l_E(\xi). \tag{13.5}$$

Since γ is compact, there is an open set V_1 such that $\gamma \subset V_1 \subset \overline{V_1} \subset V_2$. Then by modifying the metric outside of V_1, we can ensure that γ is a global minimizer of the new metric.

Proposition 13.12. *Assume that γ is a (g_E, h) (local) minimizer, and assume that the associated Hamiltonian periodic orbit η is nondegenerate. Then η is* *hyperbolic.*

Proof of Proposition 13.12. A non-degenerate periodic orbit $\eta : [0, T] \to \mathbb{T}^2 \times \mathbb{R}^2$ either is hyperbolic or the associated Poincaré map has eigenvalues on the unit circle excluding $\{1\}$. We now invoke the Green bundles introduced in Section 6.7. According to Proposition 6.19, the tangent map of the flow preserves one of the (2-dimensional) Green bundles; hence the Poincaré map preserves a 1-dimensional subspace at the periodic orbit. But this is impossible if the differential of the map has a complex eigenvalue. $\qquad\square$

Theorem 13.13. *Given $0 < e < \bar{E}$, there exists an open and dense subset $\mathcal{G}'$ of $\mathcal{G}^r$, such that for each $U \in \mathcal{G}'$, the Hamiltonian $H(\varphi, I) = K(I) - U(\varphi)$ satisfies the following statements. There exist finitely many smooth families of local minimizers*

$$\xi_j^E, \quad a_j - \sigma \leq E \leq b_j + \sigma, \, j = 1, \cdots, N,$$

and $\sigma > 0$, with the following properties.

1. *All ξ_j^E for $a_j - \sigma \leq E \leq b_j + \sigma$ are hyperbolic.*
2. *$\bigcup_j [a_j, b_j] \supset [E_0, \bar{E}]$.*
3. *For each $E_0 \leq E \leq \bar{E}$, any global minimizer is contained in the set of all ξ_j^E's such that $E \in [a_j, b_j]$.*

Proof of Theorem 13.13 occupies the rest of this section.

Lemma 13.14. *Assume that γ_{E_0} is a hyperbolic local (ρ_{E_0}, h)-minimizer. The following hold.*

1. *There exists a neighborhood V of γ_{E_0}, such that γ_{E_0} is the unique local (g_E, h)-minimizer on V.*
2. *There exists $\delta > 0$ such that for any $U' \in C^r(\mathbb{T}^2)$ with $\|U - U'\|_{C^2} \leq \delta$ and $|E' - E_0| \leq \delta$, the Hamiltonian $H'(\varphi, I) = K(I) - U'(\varphi)$ admits a hyperbolic local minimizer in V.*
3. *There exists $\delta > 0$ and a smooth family $\gamma_E \subset V$, $E_0 - \delta \leq E \leq E_0 + \delta$, such that each of them is a hyperbolic local minimizer.*

Proof. Let η_{E_0} denote the Hamiltonian orbit of γ_{E_0}. Inverse function theorem implies that if η_{E_0} is hyperbolic, then it is locally unique, and it uniquely extends to hyperbolic periodic orbit $\eta_{E,U'}$ if E is close to E_0 and U' is close to U. Let γ_E be the projection of η_E, then they must be the unique local minimizers. $\qquad\square$

We now use the information obtained to classify the set of global minimizers.

- Consider the Hamiltonian $H(\varphi, I) = K(I) - U(\varphi) + \min_\varphi U(\varphi)$. Since we only consider periodic orbits that correspond to rational homology h, the associated closed geodesic have finite length (in the Jacobi metric). For $0 < E_0 < \bar{E}$, the time change connecting the geodesic flow to the Hamiltonian system is uniformly bounded. As a result, all the global minimizers we consider have uniformed bounded period. Corollary 13.4 applies. In particular, there are only finitely many degenerate periodic orbits and the number of them is constant under small perturbation of U.

- It follows from the previous item that all but finitely many global minimizers are non-degenerate. The non-degenerate minimizers are hyperbolic by Proposition 13.12.
- Since a global minimizer is always a local minimizer, using Lemma 13.14, it extends to a smooth one parameter family of local minimizers. The extension can be continued until either the orbit is nolonger locally minimizing, or if the orbit becomes degenerate. Once a local minimizer becomes degenerate, this family can no longer be extended as local minimizers as by Proposition 13.11, it must bifurcate to a periodic orbit of complex eigenvalues.
- It is well known that for a fixed energy, any two global minimizers do not cross (see for example, [66]). We can locally extend the global minimizers for a small interval of energy without them crossing.
- There are at most finitely many families of local minimizers, because they are isolated and do not accumulate (Lemma 13.14).
- There are at most finitely many energies on which the global minimizer may be a degenerate periodic orbit.

We have proved the following statement.

Proposition 13.15. *Given $0 < e < \bar{E}$, there exists an open and dense subset $\mathcal{G}'$ of $\mathcal{G}^r$, such that for each $U \in \mathcal{G}'$, the following hold for $H(\varphi, I) = K(I) - U(\varphi) + \min_\varphi U(\varphi)$.*

1. *There are at most finitely many (maybe none) isolated global minimizers $\gamma_j^{E_j, d}$, $j = 1, \cdots, M$.*
2. *There are finitely many smooth families of local minimizers*

$$\gamma_j^E, \quad \bar{a}_j \leq E \leq \bar{b}_j, j = 1, \cdots, N,$$

   with $[e, \bar{E}] \supset \bigcup [\bar{a}_j, \bar{b}_j]$, such that γ_j^E are hyperbolic for $\bar{a}_j < E < \bar{b}_j$. The set $\bar{\gamma}_j^E$ for $E = \bar{a}_j, \bar{b}_j$ may be hyperbolic or degenerate.
3. *For a fixed energy surface E, the sets $\{\gamma_j^{E,d}\}$ and $\bigcup_{\bar{a}_j \leq E \leq \bar{b}_j} \gamma_j^E$ are pairwise disjoint.*
4. *For each $e \leq E \leq \bar{E}$, the global minimizer is chosen among the set of all $\gamma_j^{c_j, d}$ with $E = c_j$, or one of the local minimizers γ_j^E with $E \in [\bar{a}_j, \bar{b}_j]$.*

Proof of Theorem 13.13. We first show that the set of potentials U satisfying the conclusion of Theorem 13.13 is open. By Lemma 13.14, the family of local minimizers persists under small perturbations of the potential U. It suffices to show that for sufficiently small perturbation of U satisfying the conclusion, the global minimizer is still taken at one of the local families. Assume, by contradiction, that there is a sequence U_n approaching U, and for each $H_n = K - U_n$, there is some global minimizer $\gamma_n^{E_n}$ not from these families. By picking a subsequence, we can assume that it converges to a closed curve γ_*, which belong one of the local families γ_j^E. Using local uniqueness from Lemma 13.14, $\gamma_n^{E_n}$ must belong to one of the local families as well. This is a contradiction.

To prove density, it suffices to prove that for a potential U satisfying the conclusion of Proposition 13.15, we can make an arbitrarily small perturbation, such that there are no degenerate global minimizers.

Our strategy is to eliminate the degenerate global minimizers one by one using a sequence of perturbations. Suppose γ_E^d is an isolated degenerate minimizer. Then by ([28], proof of Theorem D, page 40-42), there is a perturbation δU with the property that: $\delta U \geq 0$, $U|\gamma_E^d = 0$, such that γ_E^d is still a global minimizer, but the associated Hamiltonian orbit η_E^d becomes hyperbolic (i.e. the perturbation keeps the orbit η_E^d intact but changes it's linearization). Repeating this process, we can remove all isolated degenerate minimizers.

Assume the previous step has been completed, suppose γ_E^d is the terminal point of one of the local families $\gamma_j^E : E \in [\bar{a}_j, \bar{b}_j]$, let's say $\gamma_E^d = \gamma_j^E(\bar{b}_j)$. Then since γ_j^E cannot not be extended to $E > \bar{b}_j$, the global minimizer for $E \in (\bar{b}_j, \bar{b}_j + \delta)$ must be contained in a different local branch, say $\gamma_i^E : E \in [\bar{a}_i, \bar{b}_i]$. In particular, $\gamma_i^E(\bar{b}_j)$ is another global minimizer.

Let V be a neighborhood of $\gamma_j^{E_j}$, such that $\bar{V}$ is disjoint from the set of other global minimizers with the same energy. For $\delta > 0$ we define $U_\delta : \mathbb{T}^2 \to \mathbb{R}$ such that $U_\delta|\gamma_j^{E_j} = \delta$ and $\mathrm{supp}\, U_\delta \subset V$. Let $H_\delta = K - U - U_\delta$, and let $l_{E,\delta}$ be the perturbed length function. We have

$$l_{E_j,\delta}(\gamma_j^{E_j}) = \int_{\gamma_j^{E_j}} \sqrt{2(E_j + U + \delta)K} > l_{E_j}(\gamma_j^{E_j}) = l_{E_j}(h) = l_{E_j,\delta}(h).$$

As a consequence, $\gamma_j^{E_j}$ is no longer a global minimizer for the perturbed system. Moreover, for sufficiently small δ, no new degenerate global minimizer can be created. Hence the perturbation has decreased the number of degenerate global minimizers strictly. By repeating this process finitely many times, we can eliminate all degenerate global minimizers.   □

13.3 NON-DEGENERACY AT HIGH-ENERGY

In this section we complete the proof of Theorem 4.5. This amounts to proving that finite local families of local minimizers, obtained from the previous section, are "in general position". The proof is similar to that of Proposition 3.2 (see Section 9.7).

We assume that the potential $U_0 \in \mathcal{G}^r$ satisfies the conclusions of Theorem 13.13. Let $\gamma_j^{E,U}$ denote the branches of local minimizers from Theorem 13.13, where we have made the dependence on U explicit. There exists an neighborhood $V(U_0)$ of U_0, such that the local branches $\gamma_j^{E,U}$ are defined for $E \in [a_j - \sigma/2, b_j + \sigma/2]$ and $U \in V(U_0)$.

Define a set of functions

$$f_j : [a_j - \sigma/2, b_j + \sigma/2] \times V(U_0) \to \mathbb{R}, \quad f_j(E, U) = l_E(\gamma_j^{E,U}).$$

Then $\gamma_i^{E,U}$ is a global minimizer if and only if

$$f_i(E, U) = f_{\min}(E, U) := \min_j f_j(E, U),$$

where the minimum is taken over all j's such that $E \in [a_j, b_j]$.

The following proposition implies Theorem 4.5.

Proposition 13.16. *There exists an open and dense subset V' of $V(U_0)$ such that for every $U \in V'$, the following hold:*

1. *For $E \in [E_0, \bar{E}]$, there are at most two j's such that $f_j(E, U) = f_{\min}(E, U)$.*
2. *There are at most finitely many $E \in [E_0, \bar{E}]$ such that there are two j's with $f_j(E, U) = f_{\min}(E, U)$.*
3. *For any $E \in [E_0, \bar{E}]$ and j_1, j_2 such that $f_{j_1}(E, U) = f_{j_2}(E, U) = f_{\min}(E, U)$, we have*

$$\frac{\partial}{\partial E} f_{j_1}(E, U) \neq \frac{\partial}{\partial E} f_{j_2}(E, U).$$

Proof. We claim that it suffices to prove the theorem under the additional assumption that all functions f_j's are defined on the same interval (a, b) with $f_{\min}(E, U) = \min_j f_j(E, U)$. Indeed, we may partition $[E_0, \bar{E}]$ into finitely many intervals, on which the number of local branches is constant, and prove proposition on each interval.

We define a map

$$f = (f_1, \cdots, f_N) : (a, b) \times V(U_0) \to \mathbb{R}^N,$$

and subsets

$$\Delta_{i_1, i_2, i_3} = \{(x_1, \cdots, x_n); x_{i_1} = x_{i_2} = x_{i_3}\}, \quad 1 \leq i_1 < i_2 < i_3 \leq N,$$

$$\Delta_{i_1, i_2} = \{(x_1, \cdots, x_n); x_{i_1} = x_{i_2}\}, \quad 1 \leq i_1 < i_2 \leq N$$

of $\mathbb{R}^N \times \mathbb{R}^N$. We also write $f^U(E) = f(E, U)$. The following two claims imply our proposition:

1. For an open and dense set of $U \in V(U_0)$, for all $1 \leq i_1 < i_2 < i_3 \leq N$, the set $(f^U)^{-1} \Delta_{i_1, i_2, i_3}$ is empty.
2. For an open and dense set of $U \in V(U_0)$, and all $1 \leq i_1 < i_2 \leq N$, the map $f^U : (a, b) \to \mathbb{R}^N$ is transversal to the submanifold Δ_{i_1, i_2}.

Indeed, the first claim implies the first statement of our proposition. It follows from our second claim that there are at most finitely many points in $(f^U)^{-1}\Delta_{i_1,i_2}$, which implies the second statement. Furthermore, using the second claim, we have for any $E \in (f^U)^{-1}\Delta_{i_1,i_2}$, the subspace $(Df^U(E))\mathbb{R}$ is transversal to $T\Delta_{i_1,i_2}$. This implies the third statement of our proposition.

For a fixed energy E and $(v_1, \cdots, v_N) \in \mathbb{R}^N$, let $\delta U : \mathbb{T}^2 \to \mathbb{R}$ be such that $\delta U(\varphi) = v_j$ on an open neighborhood of γ_j^E for each $j = 1, \cdots N$. Let $l_{E,\epsilon}$ and $\gamma_j^{E,\epsilon}$ denote the arclength and local minimizer corresponding to the potential $U + \epsilon \delta U$. For any φ in a neighborhood of γ_j^E, we have

$$E + U(\varphi) + \delta U(\varphi) = E + U(\varphi) + \epsilon v_j;$$

hence for sufficiently small $\epsilon > 0$, $\xi_j^{E,\epsilon} = \xi_j^{E+\epsilon v_j}$.

The directional derivative

$$D_U f(E, U) \cdot \delta U = \frac{d}{d\epsilon}\Big|_{\epsilon=0} l_{E,\epsilon}(\xi_j^{E,\epsilon}) = \frac{d}{d\epsilon}\Big|_{\epsilon=0} l_{E+\epsilon v_j}(\xi_j^{E+\epsilon v_j}) = \frac{\partial}{\partial E} f_j(E, U) v_j.$$

It follows from a direct computation that each f_j is strictly increasing in E and the derivative in E never vanishes. As a consequence, we can choose $(v_1, \cdots, v_N)$ in such a way that $D_U f(E, U) \cdot \delta U$ takes any given vector in $\mathbb{R}^N$. This implies the map

$$D_U f : (a, b) \times TV(U_0) \to \mathbb{R}^N$$

has full rank at any (E, U). As a consequence, f is transversal to any Δ_{i_1,i_2,i_3} and Δ_{i_1,i_2}. Using Sard's lemma, we obtain that for a generic U, the image of f^U is disjoint from Δ_{i_1,i_2,i_3} and f^U is transversal to Δ_{i_1,i_2}. $\qquad \square$

13.4 UNIQUE HYPERBOLIC MINIMIZER AT VERY HIGH ENERGY

In this section we prove Proposition 4.6, namely for the mechanical system

$$H^s(\varphi, I) = K(I) - U(\varphi),$$

the shortest loop in the homology $h \in H_1(\mathbb{T}^2, \mathbb{Z})$ for Jacobi metric $g_E(\varphi, v) = 2(E + U(\varphi))K^{-1}(v)$, where $K^{-1}(v)$ is the Legendre dual of K. Since the metrics

$$g_E(\varphi, v) = 2(1 + E^{-1}U(\varphi))K^{-1}(\sqrt{E}v), \quad \bar{g}_E(\varphi, v) = 2(1 + E^{-1}U(\varphi))K^{-1}(v)$$

differ by a constant multiple, it suffices to study the shortest loops for the metric $\bar{g}_E$. Denote $\delta = E^{-1}$, this is then equivalent to studying the mechanical system

$$H_\delta(\varphi, I) = K(I) - \delta U(\varphi)$$

with energy $E = 1$. Without loss of generality, we may assume $h = (1, 0)$. Let us write

$$U(\varphi) = U_1(\varphi_1) - U_2(\varphi_1, \varphi_2),$$

where $\int_0^1 U_2(\varphi_1, s)ds = 0$. We will show an averaging effect that eliminates U_2. Let us denote by ρ_0 the unique positive number such that $K^{-1}(\rho_0 h) = 1$, and let $c_0 = \partial K^{-1}(\rho_0 h)$.

Lemma 13.17. *There is $C > 1$ depending only on $\|\partial^2 K\|, \|\partial^2 K^{-1}\|$ such that then associated Hamiltonian orbit $\eta_h^1 = \{(\varphi, I)(t)\}$ satisfies*

$$\|\dot\varphi - \rho_0 h\|, \ \|I - c_0\| \le C\delta.$$

Proof. Let us note the shortest curve γ_h^1 corresponds to the orbit $\eta_h^1 = (\varphi_h, I_h)$ of the Hamiltonian H_δ, satisfying $\varphi_h(T) = \varphi_h(0) + h$. Note:

1. $\ddot\varphi_h(t) = \delta\nabla U(\varphi_h(t)) = O(\delta)$.
2. Since the energy is 1, $K(I_h(t)) = 1 + O(\delta)$, $K^{-1}(\dot\varphi_h(t)) = K^{-1}(\partial_I K(I_h(t))) = K(I_h(t)) = 1 + O(\delta)$.

The first item implies $\dot\varphi_h(t) = \frac{1}{T}h + O(\delta)$, when combined with the second item, implies $\dot\varphi_h(t) = \rho_0 h + O(\delta)$. Applying ∂K^{-1}, we get $I_h(t) = c_0 + O(\delta)$. $\square$

Note that the condition $\partial K(I) = \rho h$ corresponds precisely to the resonance segment $\Gamma = I \in \mathbb{R}^2$, $\partial K(I) \cdot (1, 0) = 0$, in other words, we are in precisely in the same situation as in single resonance. In fact, since there are fewer degrees of freedom, the normal form is much stronger.

Proposition 13.18. *Suppose $r \ge 4$. For any $M > 0$, there is $\delta_0 = \delta_0(K, M)$ and $C = C(K, M) > 0$ such that for every H_δ with $0 < \delta < \delta_0$, there is a C^∞ coordinate change Φ Hamiltonian isotopic to identity, such that*

$$H_\delta \circ \Phi = K(I) - \delta U_1(\varphi_1) - \delta R(\varphi, I),$$

with $\|R\|_{C_I^2} \le \sqrt{\delta}$, and $\|\pi_\theta(\Phi - Id)\|_{C^2} \le C\delta$ and $\|\pi_p(\Phi - Id)\|_{C^2} \le C\delta$, where the norms are measured on the set $I \in B_{M\sqrt{\delta}}(c_0)$.
If $U_1(\varphi_1)$ has a unique non-degenerate minimum at $\varphi_1 = \varphi_1^*$, and satisfies

$$U(\varphi_1) - U(\varphi_1^*) \ge \lambda\|\varphi_1 - \varphi_1^*\|^2.$$

Then by choosing $\delta_0 = \delta_0(K, M, \lambda) > 0$ smaller, the system $H_\delta \circ \Phi$ has a normally hyperbolic invariant cylinder $\mathcal{C}$ given by

$$(\varphi_1, I_1) = (F(\varphi_2, I_2) + \varphi_1^*, G(\varphi_2, I_2)), \quad \varphi_2 \in \mathbb{T}, I_2 \in \Gamma \cap B_{M\sqrt{\delta}}(c_0),$$

satisfying $\|F\|_{C^0} \le C\kappa$, $\|G\|_{C^0} \le C\kappa\sqrt{\delta}$. $\mathcal{C}$ contains all the invariant sets in

$$\|\varphi_1 - \varphi_1^*\| < C^{-1}, \quad \text{dist}(I, \Gamma \cap B_{M\sqrt{\delta}}) < C^{-1}.$$

The minimal periodic orbit η_h^1 is contained in $\mathcal{C}$ and is a graph over φ_2. It is therefore unique and hyperbolic.

Proof. The first part is very similar to Theorem 3.4, but in the much simpler case of 2-degrees of freedom. In this case, we can give out the coordinate change explicitly. Define

$$W(\varphi_1, \varphi_2) = \rho_0^{-1} \int_0^1 s U_2(\varphi_1, \varphi_2 + s) ds,$$

and note that W is well defined on $\mathbb{T}^2$ due to $\int_0^1 U_2(\varphi_1, s) ds = 0$. Note that $\partial_{\varphi_2} W = \rho_0^{-1} U_2$. Consider the coordinate change $I \mapsto I + \delta \nabla W$. Then the new Hamiltonian is

$$K(I + \delta \nabla W) - \delta U_1 - \delta U_2$$
$$= K(I) + \delta \partial K(I) \cdot \nabla W - \delta U_1 - \delta U_2 + \delta^2 K(\nabla W)$$
$$= K(I) - \delta U_1 + \delta(\rho_0 h \cdot \nabla W - U_2) + \delta^2 K(\nabla W) + \delta(\partial K(I) - \rho_0 h) \cdot \nabla W.$$

The main observation is that the second term in the last equality vanishes, the third term is $O(\delta^2)$ in C^2 norm, and the last term is $O(\delta^{\frac{3}{2}})$ in C_I^2 norm.

The other parts of the proposition is identical to Theorems 9.1 9.2, and are omitted.   □

Proposition 4.6 follows immdediately from the above proposition.

13.5   GENERIC PROPERTIES AT THE CRITICAL ENERGY

We prove Proposition 4.7 that the conditions $[DR1^c]$–$[DR4^c]$ holds on an open and dense set of U. Since all conditions are C^2 open, it suffices to prove denseness.

First, we perturb U near the origin to achieve properties $[DR1^c]$. Let W' be a ρ-neighborhood of the origin in $\mathbb{R}^2$ for small enough $\rho > 0$ so that it does not intersect sections $\Sigma_\pm^s$ and $\Sigma_\pm^u$. Consider $\xi(\varphi)$ a C^∞-bump function so that $\xi(\varphi) \equiv 1$ for $|\varphi| < \rho/2$ and $\xi(\varphi) \equiv 0$ for $|\varphi| > \rho$. Let $Q_\zeta(\varphi) = \sum \zeta_{ij} \varphi_i \varphi_j$ be a symmetric quadratic form. Consider $U_\zeta(\varphi) = U(\varphi) + \xi(\varphi)(Q_\zeta(\theta) + \zeta_0)$. In $W' \times \mathbb{R}^2$ we can diagonalize both: the quadratic form $K(p) = \langle Ap, p \rangle$ and the Hessian $\partial^2 U(0)$. Explicit calculation shows that choosing properly ζ one can make the minimum of U at 0 being unique and eigenvalues to be distinct.

Note that minimizing loops of the Jacobi metric may not intersect except possibly at 0, any two minimizing loops must be isolated away from a neighborhood of 0. We can apply localized perturbation to one of the geodesics to reduce its length, making it the global minimizer. Since the minimal loop is generically unique, its type is also determined uniquely. This proves $[DR2^c]$.

Denote $w^u = W^u \cap \Sigma_+^u$ an unstable curve on the exit section Σ_+^u and $w^s = W^s \cap \Sigma_+^s$ a stable curve on the enter section Σ_+^s. Denote on w^u (resp. w^s) the point of intersection Σ_+^u (resp. Σ_+^s) with strong stable direction q^{ss} (resp. q^{uu}). Recall that q^+ (resp. p^+) denotes the point of intersection of γ^+ with Σ_+^u (resp. Σ_+^s). $[DR3^c]$ is equivalent to conditions

$$\Phi_{\text{glob}}^+ w^u \cap w^s \neq q^{ss}, \quad (\Phi_{\text{glob}}^+)^{-1}(w^s) \cap w^u \neq q^{uu}.$$

By Lemma 13.7 the differential map $D_U \Phi$ generates a subspace orthogonal to the gradient ∇H^U and the Hamiltonian vector field $\chi^U(\cdot)$ of H^U. This means we can perturb the global map Φ_{glob}^+ to move the intersection $\Phi_{\text{glob}}^+ w^u \cap w^s$ away from q^{ss}, and similarly for q^{uu}. $[DR3^c]$ follows.

Finally, a simple critical closed loop corresponds to a homoclinic orbit η contained in the intersection $W^u \cap W^s$ of the unstable and stable manifold of the equilibrium $(0,0)$. Moreover, if we set $W_a^{s/u} = \{z \in W^{s/u} : \text{dist}(z, (0,0)) < a\}$, where dist is the distance on the surface, then there exists $a > 0$ such that $\eta \subset W_a^u \cap W_a^s$. As part of the Kupka-Smale theorem (Lemma 9 of [69]), this intersection is transversal for an open and dense set of U. This proves $[DR4^c]$ in the simple critical and non-simple case. In the simple critical case, η is a regular periodic orbit, and the statement again follows from the Kupka-Smale theorem.

Chapter Fourteen

Derivation of the slow mechanical system

In this chapter we derive the slow mechanical system at a maximal resonance. The discussions here applies to arbitrary degrees of freedom, and indeed we will consider

$$H_\epsilon(\theta, p, t) = H_0(p) + \epsilon H_1(\theta, p, t), \quad \theta \in \mathbb{T}^n, p \in \mathbb{R}^n, t \in \mathbb{T},$$

and let $p_0 \in \mathbb{R}^n$ be an n-resonance, namely, there are linearly independent $k_1, \cdots, k_n \in \mathbb{Z}^{n+1}$ such that

$$k_i \cdot (\partial H_0(p_0), 1), \quad i = 1, \cdots, n.$$

Since the resonance depends only on the hyperplane containing $k_1, \cdots, k_n$, we may choose $k_1, \cdots, k_n$ to be *irreducible*, namely

$$\mathrm{Span}_{\mathbb{Z}}\{k_1, \cdots, k_n\} = \mathrm{Span}_{\mathbb{R}}\{k_1, \cdots, k_n\} \cap \mathbb{Z}^{n+1}.$$

Results in this chapter apply to the proof of our main theorem by restricting to the case $n = 2$.

First, we will reduce the system near an n-resonance to a normal form. After that, we perform a coordinate change on the extended phase space, and an energy reduction to reveal the slow system.

In Section 14.1, we describe a resonant normal form.

In Section 14.2, we describe the affine coordinate change and the rescaling, revealing the slow system.

In Section 14.3 we discuss variational properties of these coordinate changes.

14.1 NORMAL FORMS NEAR MAXIMAL RESONANCES

Write $\omega_0 := \partial_p H_0(p_0)$, then the orbit $(\omega_0, 1) t$ is periodic. Let

$$T = \inf_{t>0} \{t(\omega_0, 1) \in \mathbb{Z}^{n+1}\}$$

be the minimal period, then there exist a constant $T_* = T_*(k_1, \cdots, k_n) > 0$, such that $T \leq T_*(k_1, \cdots, k_n)$.

Given a function $f : \mathbb{T}^n \times \mathbb{R}^n \times \mathbb{T} \to \mathbb{R}$, we define

$$[f]_{\omega_0}(\theta, p, t) = \frac{1}{T} \int_0^T f(\theta + \omega_0 s, p, t + s) ds.$$

$[f]_{\omega_0}$ corresponds to the resonant component related to the maximal resonance. Writing $H_1(\theta, p, t) = \sum_{k \in \mathbb{Z}^{n+1}} h_k(p) e^{2\pi i k \cdot (\theta, t)}$ and letting

$$\Lambda = \mathrm{Span}_{\mathbb{Z}}\{k_1, \cdots, k_n\} \subset \mathbb{Z}^{n+1},$$

then

$$[H_1]_{\omega_0}(\theta, p, t) = \sum_{k \in \Lambda} h_k(p) e^{2\pi i k \cdot (\theta, t)}.$$

We define a rescaled differential in the action variable by

$$\partial_I f(\theta, p, t) = \sqrt{\epsilon} \partial_p f(\theta, p, t). \tag{14.1}$$

Note that the I in the subscript is only a notation and does not refer to a change of variable. Let $\| \cdot \|_{C_I^r}$ denote the C^r norm with ∂_p replaced by ∂_I. For a vector or a matrix valued function, we take the C^r (or C_I^r) norm to be the sum of the norm of all elements. Let $B_r^n(p)$ denote the r-neighborhood of p_0 in $\mathbb{R}^n$.

Theorem 14.1. *Assume that $r \geq 5$. Then for any $M_1 > 0$, there exists $\epsilon_0 = \epsilon_0 > 0$, $C_1 > 1$ depending only on $\|H_0\|_{C^r}$, $T_*(k_1, \cdots, k_n)$, n, M_1 such that for any $0 < \epsilon < \epsilon_0$, there exists a C^∞ symplectic coordinate change*

$$\Phi_\epsilon : \mathbb{T}^n \times B^n_{M_1\sqrt{\epsilon}}(p_0) \times \mathbb{T} \to \mathbb{T}^n \times B^n_{2M_1\sqrt{\epsilon}}(p_0) \times \mathbb{T},$$

which is the identity in the t component, such that

$$\begin{aligned} N_\epsilon(\theta, p, t) &:= H_\epsilon \circ \Phi_\epsilon(\theta, p, t) \\ &= H_0(p) + \epsilon Z(\theta, p, t) + \epsilon Z_1(\theta, p, t, \epsilon) + \epsilon R(\theta, p, t, \epsilon), \end{aligned}$$

where $Z = [H_1]_{\omega_0}$, $[Z_1]_{\omega_0} = Z_1$, and

$$\|Z_1\|_{C_I^2} \leq C_1 \sqrt{\epsilon}, \quad \|R\|_{C_I^2} \leq C_1 \epsilon^{\frac{3}{2}}, \tag{14.2}$$

and

$$\|\Phi_\epsilon - Id\|_{C_I^2} \leq C_1 \epsilon.$$

Moreover, Φ_ϵ admits the following extensions:

1. *Φ_ϵ can be extended to $\mathbb{T}^n \times \mathbb{R}^n \times \mathbb{T}$ such that Φ_ϵ is identity outside of $\mathbb{T}^n \times U_{4M_1\sqrt{\epsilon}}(p_0) \times \mathbb{T}$.*
2. *The extension in item 1 can be further extended to $\widetilde{\Phi}_\epsilon(\theta, p, t, E)$ on $\mathbb{T}^n \times \mathbb{R}^n \times$*

$\mathbb{T} \times \mathbb{R}$, *such that*

$$\widetilde{\Phi}_\epsilon(\theta, p, t, E) = (\Phi_\epsilon(\theta, p, t), E + \widetilde{E}(\theta, p, t)),$$

and $\widetilde{\Phi}_\epsilon$ *is exact symplectic.*

3. $\|\widetilde{\Phi}_\epsilon - Id\|_{C_I^2} \leq C_1 \epsilon$ *holds.*

Remark 14.2.

- Our normal form is related to the classical "partial averaging," see for example expansion (6.5) in [2, Section 6.1.2]. Our goal here is to obtain precise control of the norms with minimal regularity assumptions. In particular, the norm estimate of $\Phi_\epsilon - Id$ is stronger than the usual results, and is needed in the proof of Proposition 14.10.
- For readers familiar with averaging theory, we perform two steps of resonant normal form transformations, and the term Z_1 groups together the resonant components of both steps.
- It is possible to lower the regularity assumption to $r \geq 4$, and use the weaker estimate $\|R\|_{C_I^2} \leq C_1 \epsilon$. This, however, requires more technical discussions in the next few sections, and we choose to avoid it.
- Due to existence of the extension, we consider $N_\epsilon = H_\epsilon \circ \Phi_\epsilon$ as defined on all of $\mathbb{T}^n \times \mathbb{R}^n \times \mathbb{T}$; however, the normal form holds only on the local neighborhood.

The rest of this section is dedicated to proving Theorem 14.1. Denote $\Pi_\theta(\theta, p, t) = \theta$, $\Pi_p(\theta, p, t) = p$ the natural projections. For a map $\Phi : \mathbb{T}^n \times U \times \mathbb{T} \to \mathbb{T}^n \times \mathbb{R}^n \times \mathbb{T}$, which is the identity in the last component, denote $\Phi = (\Phi_\theta, \Phi_p, Id)$, where $\Phi_\theta = \Pi_\theta \circ \Phi$ and $\Phi_p = \Pi_p \circ \Phi$.

Lemma 14.3. *We have the following properties of the rescaled norm.*

1. $\|f\|_{C_I^r} \leq \|f\|_{C^r}$, $\|f\|_{C^r} \leq \epsilon^{-r/2} \|f\|_{C_I^r}$.
2. $\|\partial_\theta f\|_{C_I^{r-1}} \leq \|f\|_{C_I^r}$, $\|\partial_p f\|_{C_I^{r-1}} \leq \frac{1}{\sqrt{\epsilon}} \|f\|_{C_I^r}$.
3. *There exists* $c_{r,n} > 1$ *such that* $\|fg\|_{C_I^r} \leq c_{r,n} \|f\|_{C_I^r} \|g\|_{C_I^r}$.
4. *Let* Φ *be as before, and* Id *denote the identity map. There exists* $c_{r,n} > 1$ *such that if*

$$\max\{\|\Pi_\theta (\Phi - Id)\|_{C_I^r}, \|\Pi_p(\Phi - Id)\|_{C_I^r}/\sqrt{\epsilon}\} < 1$$

we have

$$\|f \circ \Phi\|_{C_I^r} \leq c_{r,n} \|f\|_{C_I^r}.$$

Proof. The first two conclusions follow directly from the definition. For the third conclusion, we have $\|\widetilde{f}\|_{C^r} = \|f\|_{C_I^r}$, where

$$\widetilde{f}(\theta, I) = f(\theta, \sqrt{\epsilon} I).$$

The conclusion then follows from properties of the C^r-norm.

For the fourth conclusion, we note

$$f \circ \Phi = \tilde{f} \circ \tilde{\Phi},$$

where $\tilde{f}$ is as before, and $\tilde{\Phi}(\theta, I) = (\Phi_\theta(\theta, \sqrt{\epsilon}I), \Phi_p(\theta, \sqrt{\epsilon}I)/\sqrt{\epsilon})$. Moreover, let us denote $\Psi = \Phi - Id$, and note that $\tilde{\Psi} = \tilde{\Phi} - Id$. Then there exists $c > 0$ depending only on dimension such that

$$\|D\tilde{\Phi}\|_{C^{r-1}} \le c + \|D(\tilde{\Phi} - Id)\|_{C^{r-1}} \le c + \max\{\|\Pi_\theta \tilde{\Psi}\|_{C^r}, \|\Pi_p \tilde{\Psi}\|_{C^r}\}$$
$$\le c + \max\{\|\Pi_\theta (\Phi - Id)\|_{C_I^r}, \|\Pi_p(\Phi - Id)\|_{C_I^r}/\sqrt{\epsilon}\} \le c + 1,$$

To estimate the C^r norm of composition, we use the following estimate for $f, g \in C^r$, which is a consequence of the Faa-di Bruno formula (see for example, Appendix C of [33]):

$$\|f \circ g\|_{C^r} \le C \left(\|f\|_{C^1}\|g\|_{C^r} + \|f\|_{C^r}\|g\|_{C^{r-1}}^r \right). \tag{14.3}$$

We have

$$\|f \circ \Phi\|_{C_I^r} = \|\tilde{f} \circ \tilde{\Phi}\|_{C^r} \le d_r \|\tilde{f}\|_{C^r} \left(1 + \|D\tilde{\Phi}\|_{C^{r-1}}^r \right)$$
$$\le d_r \|f\|_{C_I^r}(1 + (c+1)^r) \le c_{r,n}\|f\|_{C_I^r},$$

where $c_{r,n} = d_r(1 + (c+1)^r)$. $\qquad\square$

For $\rho > 0$, denote

$$D_\rho = \mathbb{T}^n \times U_\rho(p_0) \times \mathbb{T} \times \mathbb{R}.$$

We require another lemma for estimating the norm of a Hamiltonian coordinate change. This is an adaptation of Lemma 3.15 from [33].

Lemma 14.4. *For $0 < \rho' < \rho < \rho_0$, $r \ge 4$, $2 \le \ell \le r - 1$, $\epsilon > 0$ small enough, and a C^r Hamiltonian $G : D_{\rho_0} \to \mathbb{R}$, let Φ_t^G denote the Hamiltonian flow defined by G.*

Let $c_{r,n} > 1$ be the constant from item 4, Lemma 14.3. Assume that

$$\frac{1}{\sqrt{\epsilon}}\|G\|_{C_I^{\ell+1}} < \frac{1}{c_{r,n}\sqrt{\epsilon}} \min \left\{ \sqrt{\epsilon}, \rho - \rho' \right\},$$

then for $0 \le t \le 1$, the flow Φ_t^G is well defined from $D_{\rho'}$ to D_ρ. Moreover,

$$\|\Pi_\theta(\Phi_t^G - Id)\|_{C_I^\ell} \le c_{r,n}\|\partial_p G\|_{C_I^\ell}, \quad \|\Pi_p(\Phi_t^G - Id)\|_{C_I^\ell} \le c_{r,n}\|\partial_\theta G\|_{C_I^\ell}. \tag{14.4}$$

Proof. Define

$$A_t = \max_{0 \le \tau \le t} \left\{ \|\Pi_\theta(\Phi_\tau^G - Id)\|_{C_I^\ell}, \frac{1}{\sqrt{\epsilon}}\|\Pi_p(\Phi_\tau^G - Id)\|_{C_I^\ell} \right\}.$$

Let t_0 be the largest $t \in [0,1]$ such that the following conditions hold.

(a) $A_t \leq 1$.
(b) $\Phi^G_{-t}(D_\rho) \supset D_{\rho'}$, in other words, $\Phi^G_t : D_{\rho'} \to D_\rho$ is well defined.

We first show (14.4) holds on $0 \leq t \leq t_0$, then we show $t_0 = 1$.
    By Lemma 14.3, item 4, we have

$$\left\| \Pi_\theta \left(\Phi^G_t - Id \right) \right\|_{C^l_I} \leq \int_0^t \left\| \partial_p G \circ \Phi^G_\tau \right\|_{C^l_I} d\tau \leq c_{r,n} \|\partial_p G\|_{C^l_I}$$

and

$$\left\| \Pi_p \left(\Phi^G_t - Id \right) \right\|_{C^l_I} \leq \int_0^t \left\| \partial_\theta G \circ \Phi^G_\tau \right\|_{C^l_I} d\tau \leq c_{r,n} \|\partial_\theta G\|_{C^l_I}.$$

We now show $t_0 = 1$. By the estimates obtained,

$$A_{t_0} \leq c_{r,n} \max\{\|\partial_p G\|_{C^l_I}, \frac{1}{\sqrt{\epsilon}}\|\partial_\theta G\|_{C^l_I}\} \leq \frac{c_{r,n}}{\sqrt{\epsilon}} \|G\|_{C^{l+1}_I} < 1.$$

Moreover, we have

$$\left\| \Pi_p \left(\Phi^G_{t_0} - Id \right) \right\|_{C^l_I} \leq c_{r,n} \|\partial_\theta G\|_{C^l_I} < \sqrt{\epsilon}\frac{c_{r,n}}{\sqrt{\epsilon}} \|G\|_{C^{l+1}_I} < \rho - \rho',$$

implying $\Phi_{-t_0} D_\rho \supsetneq D_{\rho'}$. If $t_0 \neq 1$, we can extend conditions (a) and (b) to $t > t_0$, contradicting the maximality of t_0. □

We now state our main technical lemma, which is an adaptation of an inductive lemma due to Bounemoura ([19]).

Lemma 14.5. *There exists a constant $K > 2$ depending only on r such that the following hold. Assume the parameters $r \geq 4$, $\rho > 0$, $\mu > 0$ satisfy*

$$0 < \epsilon \leq \mu^2, \quad KT\mu < \min\left\{\frac{1}{4}, \frac{\rho}{2(r-2)\sqrt{\epsilon}}\right\}.$$

Assume that

$$H : \mathbb{T}^n \times U_\rho(p_0) \times \mathbb{T} \times \mathbb{R} \to \mathbb{R}, \quad H(\theta, p, t, E) = l + g_0 + f_0,$$

where $l(p, E) = (\omega_0, 1) \cdot (p, E)$ is linear, g_0, f_0 are C^r and depend only on (θ, p, t), and

$$\|g_0\|_{C^r_I} \leq \sqrt{\epsilon}\mu, \quad \|f_0\|_{C^r_I(\rho_0)} \leq \epsilon, \quad \|\partial_p f_0\|_{C^{r-1}_I} \leq \epsilon. \tag{14.5}$$

Then for $j \in \{1, \cdots, r-2\}$ and $\rho_j = \rho - j\frac{\rho}{2(r-2)} > \rho/2$, there exists a collection of C^{r-j}-symplectic maps $\Phi_j : D_{\rho_j} \to D_\rho$, of the special form

$$\Phi_j(\theta, p, t, E) = (\Theta(\theta, p, t), P(\theta, p, t), t, E + \tilde{E}(\theta, p, t)).$$

The maps Φ_j have the properties

$$\|\Pi_\theta(\Phi_j - Id)\|_{C_I^{r-j}(D_{\rho_j})} \leq jK^{j+1}(T\mu)^2,$$
$$\|\Pi_p(\Phi_j - Id)\|_{C_I^{r-j}(D_{\rho_j})} \leq jK^j(T\mu)\sqrt{\epsilon}, \tag{14.6}$$

and

$$H \circ \Phi_j = l + g_j + f_j,$$

for each $j \in \{1, \cdots, r-2\}$ satisfying $g_j = g_{j-1} + [f_{j-1}]_{\omega_0}$ and

$$\|g_j\|_{C^{r-j}(D_{\rho_j})} \leq (2 - 2^{-j})\sqrt{\epsilon}\mu, \quad \|f_j\|_{C^{r-j}(D_{\rho_j})} \leq (KT\mu)^j\, \epsilon. \tag{14.7}$$

Proof. The proof is an adaptation of the proof of Proposition 3.2 in [19]. Define

$$\chi_j = \frac{1}{T} \int_0^T s(f_j - [f_j]_{\omega_0})(\theta + \omega_0 s, p, t + s)ds, \tag{14.8}$$

$\Phi_0 = Id$, and

$$\Phi_{j+1} = \Phi_j \circ \Phi_1^{\chi_j},$$

where $\Phi_s^{\chi_j}$ is the time-s map of the Hamiltonian flow of χ_j.

Using the fact that χ_j is independent of E, we have the map $\Phi^{\chi_j} := \Phi_1^{\chi_j}$ is independent of E in the (θ, p, t) components. Furthermore, Φ^{χ_j} is the identity in the t component, and $\Pi_E \Phi^{\chi_j} - E$ is independent of E. Hence Φ_j takes the special format described in the lemma. The special form of Φ_j implies that g_j and f_j are also independent of E, allowing the induction to continue.

First of all, assuming the step j of the induction is complete, we prove the norm estimate (14.6). Using (14.7),

$$\|[f_j]_{\omega_0}\|_{C_I^{r-j}} \leq \|f_j\|_{C_I^{r-j}} \leq (KT\mu)^j\epsilon, \quad \|\chi_j\|_{C_I^{r-j}} \leq 2T\|f_j\|_{C_I^{r-j}} < 2T(KT\mu)^j\epsilon.$$

We will choose K such that $K \geq 2c_{r,n}$. Then

$$\frac{1}{\sqrt{\epsilon}}\|\chi_j\|_{C_I^{r-j}} \leq \frac{1}{c_{r,n}}(2Tc_{r,n}\sqrt{\epsilon})(KT\mu)^j \leq \frac{1}{c_{r,n}}(KT\mu)^{j+1}$$
$$< \frac{1}{c_{r,n}}\max\{\frac{1}{4}, \frac{\rho}{2\sqrt{\epsilon}(r-2)}\}, \tag{14.9}$$

therefore Lemma 14.4 applies with $\rho = \rho_j$, $\rho' = \rho_{j+1}$, $G = \chi_j$.

For $j \geq 1$, using the inductive assumption and $c_{r,n} > 1$,

$$\|\partial_p \chi_j\|_{C_I^{r-j-1}} \leq \frac{1}{\sqrt{\epsilon}}\|\chi_j\|_{C_I^{r-j}} \leq (KT\mu)^{j+1},$$

while for $j = 0$, the initial assumption on f_0 implies

$$\|\partial_p \chi_0\|_{C_I^{r-1}} \leq 2T\|\partial_p f_0\|_{C_I^{r-1}} \leq 2T\epsilon \leq 2T\mu^2 < \frac{1}{c_{r,n}}(KT\mu)^2,$$

since $T \geq 1$ and $K \geq 2c_{r,n}$. Applying Lemma 14.4, we get Φ^{χ_j} are well defined maps from $D_{\rho_{j+1}}$ to D_{ρ_j}, and

$$\|\Pi_\theta(\Phi^{\chi_j} - Id)\|_{C_I^{r-j-1}} \leq c_{r,n}\|\partial_p \chi_j\|_{C_I^{r-j-1}} \leq (KT\mu)^2,$$

$$\|\Pi_p(\Phi^{\chi_j} - Id)\|_{C_I^{r-j-1}} \leq c_{r,n}\|\partial_\theta \chi_j\|_{C_I^{r-j-1}} \leq c_{r,n}\|\chi_j\|_{C_I^{r-j}} < (KT\mu)^{j+1}\sqrt{\epsilon}$$

using (14.9). Since $\Phi_1 = \Phi_1^{\chi_0}$, we obtain (14.6) for $j = 1$.

For each $j \geq 2$, using the bound on $\Phi^{\chi_j} - Id$ above, Lemma 14.3, item 4 applies for $\Phi = \Phi^{\chi_j}$. Then

$$\|\Pi_\theta(\Phi_{j-1} - Id) \circ \Phi^{\chi_{j-1}}\|_{C_i^{r-j}} \leq c_{r,n}\|\Pi_\theta(\Phi_{j-1} - Id)\|.$$

We have

$$\|\Pi_\theta(\Phi_j - Id)\|_{C_I^{r-j}}$$
$$= \|\Pi_\theta(\Phi_{j-1} \circ \Phi^{\chi_{j-1}} - \Phi^{\chi_{j-1}} + \Phi^{\chi_{j-1}} - Id)\|_{C_I^{r-j}}$$
$$\leq \|\Pi_\theta(\Phi_{j-1} - Id) \circ \Phi^{\chi_{j-1}}\|_{C_I^{r-j}} + \|\Pi_\theta(\Phi^{\chi_{j-1}} - Id)\|_{C_I^{r-j}}$$
$$\leq c_{r,n}\|\Pi_\theta(\Phi_{j-1} - Id)\|_{C_I^{r-j}} + \|\Pi_\theta(\Phi^{\chi_{j-1}} - Id)\|_{C_I^{r-j}}$$
$$\leq \sum_{i=0}^{j-1} c_{r,n}^{j-i-1}\|\Pi_\theta(\Phi^{\chi_i} - Id)\|_{C_I^{r-i-1}} < jK^{j-1}(KT\mu)^2,$$

noting $c_{r,n} < K$ and $\|\Pi(\Phi^{\chi_j} - Id)\|_{C_r^{r-j-1}} \leq (KT\mu)^2$.

By the same reasoning, we get

$$\|\Pi_p(\Phi_j - Id)\|_{C_I^{r-j}} \leq \sum_{i=0}^{j-1} c_{r,n}^{j-i-1}\|\Pi_p(\Phi^{\chi_i} - Id)\|_{C_I^{r-i-1}} \leq jK^{j-1}(KT\mu)\sqrt{\epsilon}.$$

We now proceed with the induction in j. The inductive assumption (14.7) holds for $j = 0$ due to (14.5), which is in fact stronger. For the inductive step, define

$$g_{j+1} = g_j + [f_j]_{\omega_0}.$$

Since $\|f_j\|_{C_I^{r-j}} \leq (KT\mu)^j \epsilon < 4^{-j}\sqrt{\epsilon}\mu$, we get

$$\|g_{j+1}\|_{C_I^{r-j-1}} \leq \|g_j\|_{C_I^{r-j}} + \|f_j\|_{C_I^{r-j}} \leq (2 - 2^{-j} + 4^{-j})\sqrt{\epsilon}\mu < (2 - 2^{j+1})\sqrt{\epsilon}\mu.$$

By a standard computation,

$$f_{j+1} = \int_0^1 \{f_j^s, \chi_j\} \circ \Phi_s^{\chi_j} ds,$$

where $f_j^s = g_j + sf_j + (1-s)[f_j]_{\omega_0}$. Estimate (14.7) implies $\|f_j^s\|_{C_I^{r-j}} \leq 3\|g_j\|_{C_I^{r-j}} \leq 6\sqrt{\epsilon}\mu$. Using Lemma 14.3, items 2 and 3, there exists an absolute constant $d > 1$ such that for any $f, g \in C^l(D_\rho)$, we have

$$\|\{f, g\}\|_{C_I^l} = \|\partial_\theta f \cdot \partial_p g + \partial_p f \cdot \partial_\theta g\|_{C_I^l} \leq \frac{d}{\sqrt{\epsilon}} \|f\|_{C_I^{l+1}} \|g\|_{C_I^{l+1}}.$$

Therefore

$$\|\{f_j^s, \chi_j\}\|_{C_I^{r-j-1}} \leq \frac{d}{\sqrt{\epsilon}} \|f_j^s\|_{C_I^{r-j}} \|\chi_j\|_{C_I^{r-j}}$$

$$\leq \frac{d}{\sqrt{\epsilon}} \cdot 3\sqrt{\epsilon}\mu \cdot 2T\|f_j\|_{C_I^{r-j}} \leq (6dT\mu)(KT\mu)^j \epsilon.$$

Furthermore, by Lemma 14.3, items 4,

$$\|f_{j+1}\|_{C_I^{r-j-1}} \leq \max_{0 \leq s \leq 1} \|\{f_j^s, \chi_j\} \circ \Phi_s^{\chi_j}\|_{C_I^{r-j-1}} \leq c_{r,n} \max_{0 \leq s \leq 1} \|\{f_j^s, \chi_j\}\|_{C_I^{r-j-1}}$$

$$\leq (6c_{r,n}d)T\mu(KT\mu)^j \epsilon \leq (KT\mu)^{j+1}\epsilon$$

if we choose $K \geq 6c_{r,n}d$. The induction is complete. $\qquad\square$

Proof of Theorem 14.1. First, we show that

$$H(\theta, p, t, E) = H_\epsilon(\theta, p, t) - H_0(p_0) + E = l + g_0 + f_0,$$

where $l(p, E) = (\omega_0, 1) \cdot (p, E)$,

$$g_0(\theta, p, t) = H_0(p) - H_0(p_0) - \omega_0 \cdot p$$

and $f_0 = \epsilon H_1$.

Define $\rho = 2M_1\sqrt{\epsilon}$. We have the following estimates: $\partial_\theta g_0 = 0$,

$$\|\partial_p g_0\|_{C^0} = \|\partial_p H_0(p) - \partial_p H_0(p_0)\|_{C^0} \leq \|H_0\|_{C^2} \cdot \rho = 2\|H_0\|_{C^2} M_1\sqrt{\epsilon}$$

and $\|\partial_{p^j}^j g_0\|_{C^0} \leq \|H_0\|_{C^{1+j}}$ for all $j \geq 2$. Then for some $\tilde{C}_1 > 1$ depending on H_0, H_1,

$$\|\partial_p g_0\|_{C_I^{r-1}(D_\rho)} \leq \max_{j \geq 1}(\sqrt{\epsilon})^{j-1}\|\partial_{p^j}^j g_0\|_{C^0} \leq \tilde{C}_1 M_1\sqrt{\epsilon}.$$

On the other hand, using $\|f\|_{C_I^r} \leq \|f\|_{C^r}$, we have

$$\|f_0\|_{C_I^r(D_\rho)} \leq \widetilde{C}_1 \epsilon, \quad \|\partial_p f_0\|_{C_I^{r-1}(D_\rho)} \leq \widetilde{C}_2 \epsilon.$$

Choose $\mu = \widetilde{C}_1 M_1 \sqrt{\epsilon} \geq \sqrt{\epsilon}$, we have

$$\|g_0\|_{C_I^r} \leq \max\{\|\partial_\theta g_0\|_{C_I^{r-1}(D_\rho)}, \sqrt{\epsilon}\|\partial_p g_0\|_{C_I^{r-1}(D_\rho)}\} \leq \widetilde{C}_1 M_1 \epsilon = \sqrt{\epsilon}\mu,$$

and $\|f_0\|_{C_I^r(D_\rho)}, \|\partial_p f_0\|_{C_I^{r-1}(D_\rho)} \leq \sqrt{\epsilon}\mu$. By choosing ϵ sufficiently small, we ensure

$$KT\mu = KT\widetilde{C}_1 M_1 \sqrt{\epsilon} < \min\{\frac{1}{4}, \frac{\rho}{2(r-2)\sqrt{\epsilon}}\} = \min\{\frac{1}{4}, \frac{2\bar{E}}{2(r-2)}\}.$$

The conditions of Lemma 14.5 are satisfied with these parameters and we apply the lemma with $j = 3$. There exists a map $\Phi_3 : D_{\rho/2} \to D_\rho$,

$$\|\Phi_3 - Id\|_{C_I^2(D_{\rho/2})} \leq \max\{2K^3(T\mu)^2, 2K^2(T\mu)\sqrt{\epsilon}\} \leq \widetilde{C}_2 \epsilon$$

for some $\widetilde{C}_2 > 1$ depending on $T, \widetilde{C}_1, M_1$. Moreover,

$$H \circ \Phi_3 = l + g_0 + [f_0]_{\omega_0} + [f_1]_{\omega_0} + [f_2]_{\omega_0} + f_3,$$

with

$$\|[f_1]_{\omega_0}\|_{C_I^2(D_{\rho/2})} \leq (KT\mu)\sqrt{\epsilon}\mu \leq \widetilde{C}_3 \epsilon^{\frac{3}{2}},$$

$$\|[f_2]\|_{C_I^2(D_{\rho/2})} \leq (KT\mu)^2 \sqrt{\epsilon}\mu \leq \widetilde{C}_3 \epsilon^2,$$

$$\|[f_3]\|_{C_I^2(D_{\rho/2})} \leq (KT\mu)^3 \sqrt{\epsilon}\mu \leq \widetilde{C}_3 \epsilon^{\frac{5}{2}},$$

for some $\widetilde{C}_3$ depending on $T, \widetilde{C}_1, M_1$. Using $l + g_0 = H_0(p) + E - H_0(p_0)$, define $\epsilon Z = [f_0]_{\omega_0}$, $\epsilon Z_1 = [f_1 + f_2]_{\omega_0}$, and $\epsilon R = f_3$, we obtain

$$(H_\epsilon + E) \circ \Phi_2 = H_0 + \epsilon Z + \epsilon Z_1 + \epsilon R + E$$

with the desired estimates and constant $C_1 = \max\{\widetilde{C}_2, \widetilde{C}_3\}$. Finally, we define $\Phi_\epsilon(\theta, p, t) = \Phi_3(\theta, p, t, E)$. This is well defined since Φ_3 is independent of E.

We now use a smooth approximation technique to show the normal form Φ can be taken to be C^∞. Using standard techniques, for every $\sigma > 0$, there are $C = C(r) > 0$ such that there is C^∞ functions $H_0'(p)$ and $H_1'(\theta, p, t)$ satisfying

$$\|H_0 - H_0'\|_{C^2} \leq \sigma, \quad \|H_1 - H_1'\|_{C^2} \leq \sigma, \quad \|H_0\|_{C^r}, \|H_1\|_{C^r} \leq C.$$

Note that the estimates of the normal form depend on the C^r norm of the smooth approximation. We apply the above procedure to $H_0' + \epsilon H_1'$, and obtain a C^∞ coordinate change Φ' such that

$$(H'_0 + \epsilon H'_1 + E) \circ \Phi' = H_0 + \epsilon Z' + \epsilon Z'_1 + \epsilon R'.$$

We have $\|\epsilon Z' - \epsilon Z\|_{C^2} \leq \epsilon \|H_1 - H'_1\|_{C^2} < \epsilon\sigma$ and $\|Z'_1\|_{C^2_I} \leq \tilde{C}\epsilon$, $\|R\|_{C^2_I} \leq \tilde{C}\epsilon^{\frac{3}{2}}$. We now compute

$$(H_0 + \epsilon H_1 + E) \circ \Phi'$$
$$= H_0 + \epsilon Z + \epsilon Z'_1 + +\epsilon R' + (H_0 - H'_0) \circ \Phi' + \epsilon(H_1 - H'_1) \circ \Phi' + \epsilon(Z' - Z).$$

The last three terms are bounded in C^2_I norms by $\tilde{C}\sqrt{\epsilon}^{-2}\sigma$. To ensure the remainder is small we can take $\sigma = \epsilon^4$. $\qquad\square$

The normal form lemma stated here also applies to maximal resonances with long period, which then combined with the idea of Lochak ([53]), can be used to study single resonance. This is Proposition 3.6, which we prove here.

Proof of Proposition 3.6. As in the proof of Theorem 14.1, we apply the Lemma 14.5 to $l + g_0 + f_0$, where $l + g_0 = H_0 + E$ and $f_0 = \epsilon H_1$. Choose $\rho = K_1\sqrt{\epsilon}$, then

$$\|g_0\|_{C^r_I} \leq CK_1\sqrt{\epsilon}, \quad \|f_0\|_{C^r_I} \leq \epsilon, \quad \|\partial_p f_0\|_{C^{r-1}_I} \leq \epsilon$$

where C depends on $r, n, \|H_0\|_{C^r}$. Choose

$$\mu = CK_1\sqrt{\epsilon},$$

by assumption $T \leq C_1/(K_1^2\sqrt{\epsilon})$, then $KT\mu \leq (KCC_1)K_1^{-1}$. The condition of Lemma 14.5 is satisfied if K_1 is sufficiently large. We apply Lemma 14.5 with $j = 1$, the conclusion is

$$\|\Pi_\theta(\Phi - Id)\|_{C^{r-1}_I} \leq K^2(KCC_1)^2 K_1^{-2},$$
$$\|\Pi_p(\Phi - Id)\|_{C^{r-1}_I} \leq K(KCC_1)K_1^{-1}\sqrt{\epsilon}.$$

and

$$\|f_1\|_{C^{r-j}} \leq (KCC_1)K_1^{-1}\epsilon.$$

We can also apply the same smooth approximation to upgrade the coordinate change to C^∞. $\qquad\square$

14.2 AFFINE COORDINATE CHANGE, RESCALING, AND ENERGY REDUCTION

DEFINITION OF THE SLOW SYSTEM.

Recall that an n-tuple of vectors $k_1, \cdots, k_n \in \mathbb{Z}^{n+1}$ defines an irreducible lattice, if the lattice $\Lambda = \mathrm{Span}_{\mathbb{Z}}\{k_1, \cdots, k_n\}$ satisfies $\mathrm{Span}\{k_1, \cdots, k_n\} \cap \mathbb{Z}^{n+1} = \Lambda$. Let B_0 be the $n \times (n+1)$ matrix whose rows are vectors $k_1^T, \cdots, k_n^T$ from $\mathbb{Z}^{n+1}$, and let $k_{n+1} \in \mathbb{Z}^{n+1}$ be such that

$$B := \begin{bmatrix} k_1^T \\ \vdots \\ k_{n+1}^T \end{bmatrix} \in SL(n+1, \mathbb{Z}). \tag{14.10}$$

Such a k_{n+1} exists if and only if $k_1, \cdots, k_n$ are irreducible. By possibly changing the signs of the rows of B, we ensure

$$\beta := k_{n+1} \cdot (\omega_0, 1) > 0. \tag{14.11}$$

In particular, we have

$$B \begin{bmatrix} \omega_0 \\ 1 \end{bmatrix} = \begin{bmatrix} 0 \\ \beta \end{bmatrix}. \tag{14.12}$$

Letting $A_0 = \partial_{pp}^2 H_0(p_0)$, we define

$$K(I) = \frac{1}{2} \left(B_0 A_0 B_0^T \right) I \cdot I, \quad I \in \mathbb{R}^n \tag{14.13}$$

and $U(\varphi)$, $\varphi \in \mathbb{T}^n$ by the relation

$$U(k_1 \cdot (\theta, t), \cdots, k_n \cdot (\theta, t)) = -Z(\theta, p_0, t). \tag{14.14}$$

Consider the corresponding autonomous system

$$G_\epsilon(\theta, p, t, E) = N_\epsilon + E.$$

We show that G_ϵ can be reduced to

$$H_\epsilon^s(\varphi, I, \tau) = \frac{1}{\beta} \left(K(I) - U(\varphi) + \sqrt{\epsilon} P(\varphi, I, \tau, \epsilon) \right), \quad \varphi \in \mathbb{T}^n, I \in \mathbb{R}^n, \tau \in \sqrt{\epsilon}\mathbb{T}, \tag{14.15}$$

with a coordinate change and an energy reduction. See Proposition 14.8.

LINEAR AND RESCALING COORDINATE CHANGE.

Our coordinate change is a combination of a linear and a rescaling coordinate change. Namely, we have

$$(\theta, p, t, E) = \Phi_1(\varphi, p^s, s, p^f) : \quad \begin{bmatrix} \theta \\ t \end{bmatrix} = B^{-1} \begin{bmatrix} \varphi \\ s \end{bmatrix}, \quad \begin{bmatrix} p - p_0 \\ E + H_0(p_0) \end{bmatrix} = B^T \begin{bmatrix} p^s \\ p^f \end{bmatrix},$$

$$(\varphi, p^s, s, p^f) = \Phi_2(\varphi, I, \tau, F) = (\varphi, \sqrt{\epsilon}I, \tau/\sqrt{\epsilon}, \epsilon F).$$

We then have

$$\Phi_L = \Phi_1 \circ \Phi_2 : \quad (\varphi, I, \tau, F) \in \mathbb{T}^n \times \mathbb{R}^n \times \sqrt{\epsilon}\mathbb{T} \times \mathbb{R} \mapsto (\theta, p, t, E) \quad (14.16)$$

by the formula

$$\begin{bmatrix} \theta \\ t \end{bmatrix} = B^{-1} \begin{bmatrix} \varphi \\ \tau/\sqrt{\epsilon} \end{bmatrix}, \quad \begin{bmatrix} p - p_0 \\ E + H_0(p_0) \end{bmatrix} = \sqrt{\epsilon}B^T \begin{bmatrix} I \\ \sqrt{\epsilon}F \end{bmatrix}. \quad (14.17)$$

Given any $M_1 > 0$, there exists $C_B > 1$ such that

$$\{\|(I, \sqrt{\epsilon}F)\| < C_B^{-1}M_1\} \subset \Phi_L \left(\{\|(p - p_0, E + H_0(p_0))\| < \sqrt{\epsilon}M_1\} \right)$$
$$\subset \{\|(I, \sqrt{\epsilon}F)\| < C_B M_1\}. \quad (14.18)$$

Define

$$G_\epsilon^s(\varphi, I, \tau, F) = \frac{1}{\sqrt{\epsilon}} G_\epsilon \circ \Phi_L,$$

then the Hamiltonian flows of G_ϵ^s and G_ϵ are conjugate via Φ_L (with respect to the natural symplectic forms in their corresponding spaces). This can be seen from the equivalence of the following Hamiltonian systems:

$$(G_\epsilon, \quad d\theta \wedge dp + dt \wedge dE) \sim (G_\epsilon \circ \Phi_L, \quad \sqrt{\epsilon}(d\varphi \wedge dI + d\tau \wedge dF))$$
$$\sim \left(\frac{1}{\sqrt{\epsilon}} G_\epsilon \circ \Phi_L, \quad d\varphi \wedge dI + d\tau \wedge dF \right).$$

We have the following lemma.

Lemma 14.6. *Assume that* N_ϵ *satisfies* (14.2) *on* $D_{M_1\sqrt{\epsilon}}$. *Then there exist* $C_B > 1$ *and* $C_2 = C_2(H_0, C_1, M)$ *such that*

$$G_\epsilon^s = \sqrt{\epsilon} \left(K(I) - U(\varphi) + (\beta + \sqrt{\epsilon}l(I, F)) \cdot F + \sqrt{\epsilon}P_1(\varphi, I, \tau, F, \epsilon) \right), \quad (14.19)$$

where $\beta = k_{n+1} \cdot (\omega(p_0), 1)$, *and*

$$\|l\|_{C^2}, \|P_1\|_{C^2} \leq C_1,$$

with the norm taken on the set

$$\{\|(I, \sqrt{\epsilon}F)\| < C_B^{-1} M_1\}. \tag{14.20}$$

Proof of Lemma 14.6. Due to (14.18), $G_\epsilon \circ \Phi_L$ is defined on the $\{\|(I, \sqrt{\epsilon}F)\| < C_B^{-1} M_1\}$.

Consider the expansion of N_ϵ from Theorem 14.1

$$N_\epsilon(\theta, p, t) := H_\epsilon \circ \Phi_\epsilon(\theta, p, t) = H_0(p) + \epsilon Z(\theta, p, t) + \epsilon Z_1 + \epsilon R.$$

Denote

$$A = \partial^2 (H_0(p) + E)(p_0, -H_0(p_0)) = \begin{bmatrix} \partial_{pp}^2 H_0(p_0) & 0 \\ 0 & 0 \end{bmatrix},$$

where $H_0(p) + E$ is a function of p and E. Expanding $H_0(p) + E$ to the third order at $(p_0, -H_0(p_0))$, and $Z(\cdot, p)$ to the first order at p_0, we have

$$G_\epsilon = \begin{bmatrix} \omega_0 \\ 1 \end{bmatrix} \cdot \begin{bmatrix} p - p_0 \\ E + H_0(p_0) \end{bmatrix} + \frac{1}{2} A \begin{bmatrix} p \\ E \end{bmatrix} \cdot \begin{bmatrix} p \\ E \end{bmatrix} + \epsilon Z(\theta, p_0, t) + \check{H}_0 + \epsilon \check{Z}_1 + \epsilon R,$$

with

$$\check{H}_0 = H_0(p) - H_0(p_0) - \omega(p_0) \cdot (p - p_0) - \frac{1}{2} A \begin{bmatrix} p \\ E \end{bmatrix} \cdot \begin{bmatrix} p \\ E \end{bmatrix},$$

$$\check{Z}_1 = Z(\theta, p, t) - Z(\theta, p_0, t) + Z_1.$$

On the set $\|p - p_0\| \le M_1 \sqrt{\epsilon}$ and $|E + H_0(p)| \le M_1 \sqrt{\epsilon}$, for some $C = C(H_0, \tilde{C}, M_1)$, we have

$$\|\check{H}_0\|_{C_I^2} \le C \epsilon^{\frac{3}{2}}, \quad \|\check{Z}_1\|_{C_I^2} \le C \sqrt{\epsilon}, \quad \|R\|_{C_I^2} \le C \epsilon^{\frac{3}{2}}.$$

From (14.17) we have $U(\varphi_1, \cdots, \varphi_n) = -Z(\theta, p_0, t)$, and from (14.12),

$$(\omega_0, 1) \cdot (p - p_0, E + H_0(p_0)) = (0, \beta) \cdot (\sqrt{\epsilon}I, \epsilon F) = \epsilon \beta F,$$

we get

$$G_\epsilon \circ \Phi_L =$$

$$\epsilon \left(\frac{1}{2} BAB^T \begin{bmatrix} I \\ \sqrt{\epsilon}F \end{bmatrix} \cdot \begin{bmatrix} I \\ \sqrt{\epsilon}F \end{bmatrix} + \beta F - U(\varphi) + (\check{H}_0/\epsilon + \check{Z}_1 + R) \circ \Phi_L \right). \tag{14.21}$$

Finally, note $K(I) = \frac{1}{2} BAB^T \begin{bmatrix} I \\ 0 \end{bmatrix} \cdot \begin{bmatrix} I \\ 0 \end{bmatrix}$, and define

$$l(I, F) \cdot F = BAB^T \begin{bmatrix} I \\ 0 \end{bmatrix} \cdot \begin{bmatrix} 0 \\ \sqrt{\epsilon}F \end{bmatrix} + \frac{1}{2} BAB^T \begin{bmatrix} 0 \\ \sqrt{\epsilon}F \end{bmatrix} \cdot \begin{bmatrix} 0 \\ \sqrt{\epsilon}F \end{bmatrix},$$

$$\sqrt{\epsilon}P = (\check{H}_0 + \epsilon \check{Z}_1 + R) \circ \Phi_L.$$

Then (14.19) follows directly from (14.21) and definition of Φ_L. It is also clear from the definition that $l(I, F)/\sqrt{\epsilon}$ has bounded C^2 norm on the set $\{\|I\|, \|\sqrt{\epsilon}F\| \leq C_B^{-1} M\}$.

For the norm estimates, we note the coordinate change Φ_1 increases the C^2 norm by a factor C_B depending only on B. Note that $\check{Z}_1$ depends only on $k_1 \cdot (\theta, t), \cdots, k_n \cdot (\theta, t), p$, and $\check{Z}_1$ depends only on φ, I, F (and not in τ). As a result, using the definition of the rescaled norm (14.3), we have

$$\|(\check{H}_0/\epsilon + \check{Z}_1) \circ \Phi_L\|_{C^2} \leq \|(\check{H}_0/\epsilon + \check{Z}_1) \circ \Phi_1(\varphi, \sqrt{\epsilon}I, \epsilon F)\|_{C_I^2}$$
$$\leq \|(\check{H}_0/\epsilon + \check{Z}_1) \circ \Phi_1\|_{C^2} \leq C_B C \sqrt{\epsilon}.$$

For terms depending on τ, we have

$$\|R \circ \Phi_L\|_{C^2} = \|R \circ \Phi_1(\varphi, \tau/\sqrt{\epsilon}, \sqrt{\epsilon}I, \epsilon F)\|_{C^2} \leq C_B(\sqrt{\epsilon})^{-2} \|R\|_{C_I^2} \leq C_B C \sqrt{\epsilon},$$

where we used (14.2). The C^2-norm estimate of P follows from estimates on C_I^2 norms of $\check{H}_0$, $\check{Z}_1$, and R. $\qquad \square$

REDUCTION ON ENERGY SURFACE

We perform a standard reduction on the energy surface $G_\epsilon^s = 0$, with τ as the new time, obtaining a time-periodic system.

Lemma 14.7. *Assume that the conclusions of Lemma 14.6 hold on the set* $\{\|(I, \sqrt{\epsilon}F)\| < M_2\}$. *Then there exist*

$$\epsilon_0 = \epsilon_0(H_0, C_1, M_2), \quad C_3 = C_3(H_0, C_1, M_2, U) > 0,$$

such that for any $0 < \epsilon < \epsilon_0$, *there is a function*

$$H_\epsilon^s : \mathbb{T}^n \times B_{M_2}^n(0) \times \sqrt{\epsilon}\mathbb{T} \to \mathbb{R}$$

uniquely solving the equation

$$G_\epsilon^s(\varphi, I, \tau, -H_\epsilon^s) = 0$$

on the set $\{\|(I, \sqrt{\epsilon}F)\| \leq M_2\}$. *Moreover,* H_ϵ^s *has the form (14.15), i.e.*

$$H_\epsilon^s(\varphi, I, \tau) = \frac{1}{\beta} \left(K(I) - U(\varphi) + \sqrt{\epsilon}P(\varphi, I, \tau, \epsilon) \right)$$

where $\|P\|_{C^2} \leq C_3$.

In particular, $H_\epsilon^s \to H^s/\beta$ *uniformly in* $C^2(\mathbb{T}^2 \times \mathbb{R}^2 \times \mathbb{R})$.

Proof of Lemma 14.7. We can choose ϵ_0 such that for any $0 < \epsilon < \epsilon_0$

$$\frac{\partial}{\partial F}(G^s_\epsilon) > \frac{\beta}{2} > 0$$

on $\{\|(I, \sqrt{\epsilon}F)\| \le M_2\}$. Therefore, H^s_ϵ exists by the implicit function theorem. Moreover, there exists a constant C', depending on H_0 and U, but independent of ϵ, such that $\|H^s_\epsilon\|_{C^2} \le C'$.

Let $Q = H^s_\epsilon - \frac{1}{\beta}(K(I) - U(\varphi)) = H^s_\epsilon - H^s$, then $\|Q\|_{C^2} \le \|H^s\|_{C^2} + C'$. We know

$$
\begin{aligned}
0 &= G^s_\epsilon(\varphi, I, \tau, -Q - H^s) \\
&= \sqrt{\epsilon}\left(-\beta Q + \sqrt{\epsilon}l(I, Q + H^s)(Q + H^s) + \sqrt{\epsilon}P_1(\varphi, I, \tau, Q + H^s\epsilon)\right) \\
&=: \sqrt{\epsilon}(-\beta Q + \sqrt{\epsilon}P_2(\varphi, I, \tau, Q + H^s, \epsilon)).
\end{aligned}
$$

Therefore,

$$Q = \frac{\sqrt{\epsilon}}{\beta}P_2(\varphi, I, \tau, Q + H^s, \epsilon).$$

To solve this implicit equation notice that there exists $C'' > 0$ depending only on C_1, H^s such that $\|P_2\|_{C^2} \le C''$. Application of the Faa-di Bruno formula ((14.3)) shows that for some $C > 0$ depending only on n we have

$$
\begin{aligned}
\|Q\|_{C^2} &= \frac{\sqrt{\epsilon}}{\beta}\|P_2(\varphi, I, \tau, Q + H^s, \epsilon)\|_{C^2} \\
&\le \sqrt{\epsilon}CC''\|Q\|^2_{C^2} \le \sqrt{\epsilon}CC''(\|H^s\|_{C^2} + C'),
\end{aligned}
$$

and the lemma follows. □

We have

$$G^s_\epsilon(\varphi, I, \tau, F) = 0, \quad \Longleftrightarrow \quad H^s_\epsilon + F = 0. \tag{14.22}$$

The following Proposition follows from the standard energy reduction (see for example [8]).

Proposition 14.8. *Assume the conclusions of Theorem 14.1 hold on the set*

$$D_{M_1} := \{\|(p - p_0, H + H_0(p_0)\| < M_1\sqrt{\epsilon}\}$$

for sufficiently large M_1 depending only on H_0. Then for $M_3 = C_B^{-2}M_1$, $0 < \epsilon < \epsilon_0(H_0, C_1, M_1)$, the following are equivalent:

1. *The curve $(\theta, t, p, E)(t)$ is an orbit of $N_\epsilon(\theta, p, t) + E$ inside D_{M_1}.*
2. *The curve $(\varphi, \tau, I, F)(t) = \Phi_L(\theta, t, p, E)(t)$ is the time change of an orbit of $N^s_\epsilon(\varphi, I, \tau) + F$, with τ as the new time.*

14.3 VARIATIONAL PROPERTIES OF THE COORDINATE CHANGES

We have made two reductions: The normal form

$$H_\epsilon(\theta, p, t) \to N_\epsilon(\theta, p, t) = H_\epsilon \circ \Phi_\epsilon,$$

and the coordinate change with time change

$$N_\epsilon(\theta, p, t) + E \to H_\epsilon^s(\varphi, I, \tau) + F.$$

In this section, we discuss the effect of these reductions on the Lagrangian, barrier function, and the Mather, Aubry, and Mañé sets. The main conclusion of this section is the following proposition, which follows directly from Propositions 14.10 and 14.11.

Proposition 14.9. *For M_1 large enough depending only on $H_0, k_1, \cdots, k_n, n$, there exist $C_2 > 1$ and $\epsilon_0 > 0$ depending on $H_0, k_1, \cdots, k_n, n, M_1$, such that the following hold for any $0 < \epsilon < \epsilon_0$.*

1. *For $M' = C_B^{-2} M_1/2$ (see (14.20)), and $\|c - p_0\| \le M'\sqrt{\epsilon}$ and $\alpha = \alpha_{H_\epsilon}(c)$, let $\bar{c}$ and $\bar{\alpha}$ satisfy*

$$\begin{bmatrix} c - p_0 \\ -\alpha + H_0(p_0) \end{bmatrix} = B^T \begin{bmatrix} \sqrt{\epsilon}\,\bar{c} \\ -\epsilon\,\bar{\alpha} \end{bmatrix}$$

   then $\alpha_{H_\epsilon^s}(\bar{c}) = \bar{\alpha}$.
2. *Let $c, \alpha_{N_\epsilon}(c), \bar{c}, \alpha_{H_\epsilon^s}(\bar{c})$ satisfy the relation in item 1, and suppose $(\varphi_i, t_i) \in \mathbb{T}^n \times \mathbb{T}, (\varphi_i, \tau_i) \in \mathbb{T}^n \times \sqrt{\epsilon}\,\mathbb{T}, i = 1, 2$ satisfies*

$$\begin{bmatrix} \theta_i \\ t_i \end{bmatrix} = B^{-1} \begin{bmatrix} \varphi_i \\ \tau_i/\sqrt{\epsilon} \end{bmatrix} \quad \mod \mathbb{Z}^n \times \mathbb{Z}$$

   then

$$|h_{H_\epsilon, c}(\theta_1, t_1; \theta_2, t_2) - \sqrt{\epsilon} h_{H_\epsilon^s, \bar{c}}(\varphi_1, \tau_1; \varphi_2, \tau_2)| \le C\epsilon.$$

3. *Let $c, \alpha_{H_\epsilon}(c), \bar{c}, \alpha_{H_\epsilon^s}(\bar{c})$ be as before; then*

$$\widetilde{\mathcal{M}}_{H_\epsilon^s}(\bar{c}) = \Phi_L \circ \Phi_\epsilon(\widetilde{\mathcal{M}}_{H_\epsilon}(c)), \quad \widetilde{\mathcal{A}}_{H_\epsilon^s}(\bar{c}) = \Phi_L \circ \Phi_\epsilon(\widetilde{\mathcal{A}}_{H_\epsilon}(c)),$$

$$\widetilde{\mathcal{N}}_{H_\epsilon^s}(\bar{c}) = \Phi_L \circ \Phi_\epsilon(\widetilde{\mathcal{N}}_{H_\epsilon}(c)).$$

Recall that Φ_ϵ can be extended to the whole phase space, and $N_\epsilon = H_\epsilon \circ \Phi_\epsilon$ is considered as a function on $\mathbb{T}^n \times \mathbb{R}^n \times \mathbb{T}$.

Proposition 14.10. *In the setup of 14.9 there exists $C_3 > 1$ depending on $H_0, k_1, \cdots, k_n, n, M_1$ such that we have the following relation.*

1. *$\alpha_{H_\epsilon}(c) = \alpha_{N_\epsilon}(c)$, $\Phi_\epsilon \widetilde{\mathcal{M}}_{H_\epsilon}(c) = \widetilde{\mathcal{M}}_{N_\epsilon}(c)$, $\Phi_\epsilon \widetilde{\mathcal{A}}_{H_\epsilon}(c) = \widetilde{\mathcal{A}}_{N_\epsilon}(c)$, $\Phi_\epsilon \widetilde{\mathcal{N}}_{H_\epsilon}(c) = \widetilde{\mathcal{N}}_{N_\epsilon}(c)$.*

2. $|A_{H_\epsilon,c}(\theta_1,\tilde{t}_1;\theta_2,\tilde{t}_2) - A_{N_\epsilon,c}(\theta_1,\tilde{t}_1;\theta_2,\tilde{t}_2)| \le C_3\epsilon.$

3. $|h_{H_\epsilon,c}(\theta_1,t_1;\theta_2,t_2) - h_{N_\epsilon,c}(\theta_1,t_1;\theta_2,t_2)| \le C_3\epsilon.$

Proof of Proposition 14.10. The symplectic invariance of the alpha function and the Mather, Aubry and Mañé sets follows from the fact that $\tilde{\Phi}_\epsilon$ is exact and Hamiltonian isotopic to identity (see [10] and Proposition 2.7). In order to get the quantitative estimate of action, we need more detailed estimates.

Writing $\tilde{\Phi}_\epsilon(\theta, p, t, E) = (\Theta, P, t, \Phi_E)$, from Theorem 14.1 and using the rescaled norm (14.3), we have

$$\|\tilde{\Phi}_\epsilon - Id\|_{C^0} \le C_1\epsilon, \quad \|\Phi_\epsilon - Id\|_{C^1} \le C_1\sqrt{\epsilon}$$

Denote $\tilde{E} = \Phi_E - E$. By exactness of $\tilde{\Phi}_\epsilon$, we have there exists a function $S : \mathbb{T}^n \times \mathbb{R}^n \times \mathbb{T} \times \mathbb{R} \to \mathbb{R}$ such that

$$Pd\Theta + \Phi_E dt - (pd\theta + Edt) = Pd\Theta - pd\theta + \tilde{E}dt = dS(\theta, p, t, E) = dS(\theta, p, t) \tag{14.23}$$

In particular, given a curve $(\theta, p, t, E)(t)$, $t \in [\tilde{t}_1, \tilde{t}_2]$ with $N_\epsilon + E = 0$, we have for $(\Theta, P, t, \Phi_E)(t) = \tilde{\Phi}_\epsilon(\theta, p, t, E)$ (and hence $H_\epsilon(\Theta, P, t) + \Phi_E = 0$), and we apply (14.23) to the tangent vector of the curve to get

$$\frac{d}{dt}S(\theta, p, t, E) = P \cdot \dot{\Theta} + \Phi_E - (p \cdot \dot{\theta} + E)$$
$$= P \cdot \dot{\Theta} - H_\epsilon(\Theta, P, t) - (p \cdot \dot{\theta} - N_\epsilon(\theta, p, t))$$
$$= L_{H_\epsilon}(\Theta, \dot{\Theta}, t) - L_{N_\epsilon}(\theta, \dot{\theta}, t).$$

As a result,

$$\int_{\tilde{t}_1}^{\tilde{t}_2} \left(L_{H_\epsilon}(\Theta, \dot{\Theta}, t) - c \cdot \dot{\Theta}\right) dt - \int_{\tilde{t}_1}^{\tilde{t}_2} \left(L_{N_\epsilon}(\theta, \dot{\theta}, t) - c \cdot \dot{\theta}\right) dt$$
$$= \left(S(\theta, p, t, E) - c \cdot (\tilde{\Theta} - \tilde{\theta})\right)\Big|_{\tilde{t}_1}^{\tilde{t}_2}, \tag{14.24}$$

where $\tilde{\Theta}, \tilde{\theta}$ are lifts of $\Theta(t), \theta(t)$ to the universal cover. Moreover, from

$$\|\Phi_\epsilon - Id\|_{C^0} \le C_1\epsilon,$$

we get $\Theta - \theta$ is a well defined vector function on $\mathbb{T}^n \times B_{M_1}(p_0)^n \times \mathbb{T}$. In particular, we have $\tilde{\Theta} - \tilde{\theta} = \Theta - \theta$.

We now estimate the C^0-norm of S. Write $S_0 = p \cdot (\Theta - \theta)$, using (14.23) we have

$$dS = (P - p)d\Theta + pd(\Theta - \theta) + \tilde{E}dt = (P - p)d\Theta + dS_0 - (\Theta - \theta)dp + \tilde{E}dt.$$

Since $\|\Theta - \theta\|_{C^0}, \|P - p\|_{C^0}, \|\tilde{E}\|_{C^0} = \|\Phi_E - E\|_{C^0} \le C_1\epsilon$, we have $\|dS\|_{C^0} \le C_1\epsilon$, and $\|S - S(0)\|_{C^0} \le C'\epsilon$ for some C' depending on C_1.

It follows that the integral in (14.24) is bounded by $C'\epsilon$. Applying this estimate to a minimizer, we have

$$|A_{N_\epsilon,c}(\theta(\tilde{t}_1), \tilde{t}_1; \theta(\tilde{t}_2), \tilde{t}_2) - A_{H_\epsilon,c}(\Theta(\tilde{t}_1), \tilde{t}_1; \Theta(\tilde{t}_2), \tilde{t}_2))| \le C'\epsilon.$$

Since $\|\Phi_\epsilon - Id\|_{C^0} \le C_1\epsilon$, we have $\|\theta(\tilde{t}_i) - \Theta(\tilde{t}_i)\| \le C_1\epsilon$, $i = 1, 2$. The estimate follows from the Lipschitz property of $A_{H,c}$.

Taking the limit, we obtain the estimate for the barrier $h_{H,c}$. $\qquad\square$

We now study the relation between N_ϵ and H_ϵ^s. Extend the definition of H_ϵ^s to $\mathbb{T}^n \times \mathbb{R}^n \times \mathbb{R}$ so that the $\sqrt{\epsilon}P$ term is supported on the set $\{\|(I, \sqrt{\epsilon}F)\| < 2M_2\}$.

Proposition 14.11. *Assume the conclusions of Theorem 14.1 hold on the set*

$$\{\|(p - p_0, H + H_0(p_0)\| < M_1\sqrt{\epsilon}\}$$

for sufficently large M_1 depending only on H_0. Then for $M_3 = C_B^{-2}M_1$, we have:

1. *Let (θ, p, t, E) satisfies $\|(p - p_0, E + H_0(p_0))\| < M_3$, $N_\epsilon(\theta, p, t) + E = 0$, and*

$$(\varphi, I, \tau, F) = \Phi_L(\theta, p, t, E).$$

Let L_{N_ϵ} and $L_{H_\epsilon^s}$ be the Lagrangians for N_ϵ and H_ϵ^s. Then for

$$v = \partial_p N_\epsilon(\theta, p, t), \qquad \begin{bmatrix} v^s \\ v^f \end{bmatrix} = B \begin{bmatrix} v \\ 1 \end{bmatrix},$$

we have

$$L_{N_\epsilon}(\theta, v, t) - p_0 \cdot v + H_0(p_0) = \epsilon v^f L_{H_\epsilon^s} \left(\varphi, \frac{1}{\sqrt{\epsilon}} \frac{v^s}{v^f}, s\sqrt{\epsilon} \right).$$

2. *Suppose $(c, \alpha), (\bar{c}, \bar{\alpha}) \in \mathbb{R}^n \times \mathbb{R}$ satisfies $\|c - p_0\| \le \frac{M_3}{2}\sqrt{\epsilon}$ and*

$$\begin{bmatrix} c - p_0 \\ -\alpha + H_0(p_0) \end{bmatrix} = B^T \begin{bmatrix} \sqrt{\epsilon}\bar{c} \\ \epsilon\bar{\alpha} \end{bmatrix},$$

then $\alpha = \alpha_{N_\epsilon}(c)$ if and only if $\bar{\alpha} = \alpha_{H_\epsilon^s}(\bar{c})$.

3. *Let $c, \alpha_{N_\epsilon}(c), \bar{c}, \alpha_{H_\epsilon^s}(\bar{c})$ satisfies the relation in item 2, and suppose $(\varphi_i, t_i) \in \mathbb{T}^n \times \mathbb{T}$, $(\varphi_i, \tau_i) \in \mathbb{T}^n \times \sqrt{\epsilon}\mathbb{T}$, $i = 1, 2$ satisfies*

$$\begin{bmatrix} \theta_i \\ t_i \end{bmatrix} = B^{-1} \begin{bmatrix} \varphi_i \\ \tau_i/\sqrt{\epsilon} \end{bmatrix} \qquad \text{mod } \mathbb{Z}^n \times \mathbb{Z},$$

then

$$h_{N_\epsilon,c}(\theta_1, t_1; \theta_2, t_2) = \sqrt{\epsilon}h_{H_\epsilon^s,\bar{c}}(\varphi_1, \tau_1; \varphi_2, \tau_2).$$

4. *Let $c, \alpha_{N_\epsilon}(c), \bar{c}, \alpha_{H_\epsilon^s}(\bar{c})$ be as before, then*

$$\widetilde{\mathcal{M}}_{H_\epsilon^s}(\bar{c}) = \Phi_L(\widetilde{\mathcal{M}}_{N_\epsilon}(c)), \quad \widetilde{\mathcal{A}}_{H_\epsilon^s}(\bar{c}) = \Phi_L(\widetilde{\mathcal{A}}_{N_\epsilon}(c)), \quad \widetilde{\mathcal{N}}_{H_\epsilon^s}(\bar{c}) = \Phi_L(\widetilde{\mathcal{N}}_{N_\epsilon}(c)).$$

Proof of Proposition 14.11. The choice of M_3 is to ensure that for $M_2 = C_B^{-1} M_1$ as in Lemma 14.6,

$$\Phi_L(\{\|(p - p_0, E + H_0(p_0))\| < M_3\}) \subset \{\|(I, \sqrt{\epsilon}F)\| < M_2\}.$$

Item 1. Recall that $G_\epsilon = H_\epsilon + E$, and

$$L_{N_\epsilon} - p_0 \cdot v + H_0(p_0) = \sup_p \{(p - p_0) \cdot v - N_\epsilon(\theta, p, t) + H_0(p_0)\}$$
$$= \sup_{p, E} \{(p - p_0, E + H_0(p_0)) \cdot (v, 1) : \quad N_\epsilon(\theta, p, t) + E = 0\}, \quad (14.25)$$

where the supremum is acheived at a unique point since $H_\epsilon + E \leq 0$ is a strictly convex set. Let us denote $L_{N_\epsilon}^{p_0} = L_{N_\epsilon} - p_0 \cdot v + H_0(p_0)$.

Continuing from (14.25) and using (14.17), we get

$$L_{N_\epsilon}^{p_0}(\theta, v, t) = \sup \left\{ (B^T)^{-1} \begin{bmatrix} p - p_0 \\ E + H_0(p_0) \end{bmatrix} \cdot B \begin{bmatrix} v \\ 1 \end{bmatrix} : \quad G_\epsilon(\theta, p, t, E) = 0 \right\}$$
$$= \sup \{ \sqrt{\epsilon}(I, \sqrt{\epsilon}F) \cdot (v^s, v^f) : \quad G_\epsilon^s(\varphi, I, \tau, F) = 0 \}$$
$$= \sqrt{\epsilon} \sup \{ (I, F) \cdot (v^s, v^f \sqrt{\epsilon}) : \quad H_\epsilon^s(\varphi, I, \tau) + F = 0 \},$$
$$= \epsilon v^f \sup \{ (I, F) \cdot (v^s/(v^f \sqrt{\epsilon}), 1) : \quad H_\epsilon^s(\varphi, I, \tau) + F = 0 \}$$
$$= \epsilon v^f L_{H_\epsilon^s}(\varphi, v^s/(v^f \sqrt{\epsilon}), \tau).$$

Item 2. We first derive a relation between the integrals of the Lagrangians. Let $(\theta, p, t, E)(t), t \in [t_1, t_2]$ be a solution to $N_\epsilon + E$ with

$$N_\epsilon + E = 0 \quad \text{and} \quad \|(p - p_0, E + H_0(p_0))\| < M_1 \sqrt{\epsilon}.$$

Let $(\varphi, I, \tau, F)(t) = \Phi_L^{-1}(\theta, p, t, E)(t)$. Then from (14.17),

$$\begin{bmatrix} \dot{\varphi} \\ \dot{\tau}/\sqrt{\epsilon} \end{bmatrix} = B \begin{bmatrix} \dot{\theta} \\ 1 \end{bmatrix} := \begin{bmatrix} v^s \\ v^f \end{bmatrix}(t), \quad \frac{d\varphi}{d\tau} = \frac{v^s}{\sqrt{\epsilon}v^f}.$$

From item 1 we get

$$\int_{t_1}^{t_2} L_{N_\epsilon}^{p_0}(\theta, \dot{\theta}, t) dt = \int_{t_1}^{t_2} \epsilon v^f(t) L_{H_\epsilon^s}\left(\varphi, v^s/(v^f \sqrt{\epsilon}), \tau\right) dt$$
$$= \int_{\tau_1}^{\tau_2} \epsilon v^f(\tau) L_{H_\epsilon^s}(\varphi(\tau), \frac{d\varphi}{d\tau}(\tau), \tau) \frac{dt}{d\tau}(\tau) d\tau = \sqrt{\epsilon} \int_{\tau_1}^{\tau_2} L_{H_\epsilon^s}(\varphi, \frac{d\varphi}{d\tau}, \tau) d\tau.$$

Let $\widetilde{\theta}(t)$ be a lift of $\theta(t)$ to the universal cover and $\widetilde{\varphi}(t), \widetilde{\tau}(t)$ be a lift of

$(\varphi(t), \tau(t))$. Then

$$\begin{bmatrix} \widetilde{\varphi}(t) \\ \widetilde{\tau}(t)/\sqrt{\epsilon} \end{bmatrix} = B \begin{bmatrix} \widetilde{\theta}(t) \\ t \end{bmatrix} + const.$$

Given any $(c, -\alpha) \in \mathbb{R}^n \times \mathbb{R}$, we have

$$\int_{t_1}^{t_2} \left(L_{N_\epsilon}(\theta, \dot{\theta}, t) - c \cdot \dot{\theta} + \alpha \right) dt$$

$$= \int_{t_1}^{t_2} L_{N_\epsilon}^{p_0}(\theta, \dot{\theta}, t) dt - (c - p_0, -\alpha + H_0(p_0)) \cdot \left((\widetilde{\theta}(t_2) - \widetilde{\theta}(t_1), t_2 - t_1) \right)$$

$$= \sqrt{\epsilon} \int_{\tau_1}^{\tau_2} L_{H_\epsilon^s} d\tau - (B^T)^{-1} \begin{bmatrix} c - p_0 \\ -\alpha + H_0(p_0) \end{bmatrix} \cdot B \begin{bmatrix} \widetilde{\theta}(t_2) - \widetilde{\theta}(t_1) \\ t_2 - t_1 \end{bmatrix}$$

$$= \sqrt{\epsilon} \int_{\tau_1}^{\tau_2} L_{H_\epsilon^s} d\tau - (\sqrt{\epsilon}\bar{c}, -\epsilon\bar{\alpha}) \cdot (\widetilde{\varphi}(\tau_2) - \widetilde{\varphi}(\tau_1), (\widetilde{\tau}_2 - \widetilde{\tau}_1)/\sqrt{\epsilon})$$

$$= \sqrt{\epsilon} \int_{\tau_1}^{\tau_2} \left(L_{H_\epsilon^s} - \bar{c} \cdot \frac{d\varphi}{d\tau} + \bar{\alpha} \right) (\varphi, \frac{d\varphi}{d\tau}, \tau) d\tau.$$

$$(14.26)$$

We now prove item 2. Suppose $\alpha = \alpha_{N_\epsilon}(c)$, let $(\theta, p, t, E)(t)$, $t \in [-\infty, \infty)$ be an orbit in the Mañé set $\mathcal{N}_{N_\epsilon}(c)$. Then there exists $C_* > 0$ such that we have $\|p(t) - c\| \le C^* \sqrt{\epsilon}$, by Proposition 7.6. Since $\|c - p_0\| \le M_3 \sqrt{\epsilon}/2$, with M_3 large enough, we have $\|p(t) - p_0\| \le M_3 \sqrt{\epsilon}$ for all t.

Moreover, using the fact that the orbit is semi-static, we have

$$-C \le \int_0^T (L_{N_\epsilon} - c \cdot v + \alpha)(\theta(t), \dot{\theta}(t), t) dt \le C.$$

From (14.26), we get

$$-C \le \int_{\tau(0)}^{\tau(T)} \left(L_{H_\epsilon^s} - \bar{c} \cdot v + \bar{\alpha} \right) d\tau \le C$$

for all $T > 0$. This implies $\bar{\alpha}$ is Mañé critical for $\bar{c}$.

As a result, we obtain that α is Mañé critical for $L_{N_\epsilon} - c \cdot v$ implies $\bar{\alpha}$ is Mañé critical for $L_{H_\epsilon^s} - \bar{c} \cdot v$. Moreover, the converse is true by reversing the above computations.

Item 3. Apply (14.26) to any one-sided minimizer of $h_{N_\epsilon, c}(\theta_1, t_1; \theta_2, t_2)$, and use the localization of calibrated orbits.

Item 4. (14.26) implies that an orbit $(\theta, p, t, E)(t)$ is semi-static for $L_{N_\epsilon, c}$ implies $(\varphi, I, \tau, F) = \Phi_L^{-1}(\theta, p, t, E)$ is a reparametrization of a semi-static orbit of $L_{H_\epsilon^s, \bar{c}}$. The converse also holds. This implies the relation between Mañé sets. The same applies to Aubry sets. Note that the Mather set is precisely the closure of the union of the support of all invariant measures contained in the Aubry set, and therefore is also invariant. □

Chapter Fifteen

Variational aspects of the slow mechanical system

In this chapter we study the variational properties of the slow mechanical system

$$H^s(\varphi, I) = K(I) - U(\varphi),$$

with $\min U = U(0) = 0$.

The main goal of this section is to derive some properties of the "channel" $\bigcup_{E>0} \mathcal{LF}_\beta(\lambda_h^E h)$, and information about the Aubry sets for $c \in \mathcal{LF}_\beta(\lambda_h^E h)$. More precisely, we prove Proposition 5.1 and justify the picture in Figure 5.1.

- In Section 15.1, we show that each $\mathcal{LF}_\beta(\lambda_h^E h)$ is a segment parallel to $h^\perp$.
- In Section 15.2, we provide a characterization of the segment and provide information about the Aubry sets.
- In Section 15.3, we provide a condition for the "width" of the channel to be non-zero.
- In Section 15.4, we discuss the limit of the set $\mathcal{LF}_\beta(\lambda_h^E h)$ as $E \to 0$, which corresponds to the "bottom" of the channel.

We drop all superscripts "s" to simplify the notations. The results proved in this section are mostly contained in [64] in some form. Here we reformulate some of them for our purpose and also provide some different proofs.

15.1 RELATION BETWEEN THE MINIMAL GEODESICS AND THE AUBRY SETS

Assume that $H(\varphi, I)$ satisfies the conditions $[DR1^h]$–$[DR3^h]$ and $[DR1^c]$–$[DR4^c]$. Then for $E \neq E_j$, $1 \leq j \leq N - 1$, there exists a unique shortest geodesic γ_h^E for the metric g_E in the homology h. For the bifurcation values $E = E_j$, there are two shortest geodesics γ_h^E and $\bar{\gamma}_h^E$.

The function $l_E(h)$ denotes the length of the shortest g_E-geodesic in homology h. By Lemma 15.4 the length function $l_E(h)$ is continuous and strictly increasing in $E \geq 0$.

Assume that the curves γ_h^E are parametrized using the Maupertuis principle, namely, it is the projection of the associated Hamiltonian orbit. Let $T(\gamma_h^E)$ be the period under this parametrization, and write $\lambda(\gamma_h^E) = 1/(T(\gamma_h^E))$.

We pick another vector $\bar{h} \in H_1(\mathbb{T}^2, \mathbb{Z})$ such that $h, \bar{h}$ form a basis of $H_1(\mathbb{T}^2, \mathbb{Z})$

and for the dual basis $h^*, \bar{h}^*$ in $H^1(\mathbb{T}^2, \mathbb{R})$ we have $\langle h, \bar{h}^* \rangle = 0$. We denote $\bar{h}^*$ by $h^\perp$ to emphasize the latter fact.

Items 1–4 of the next theorem will be proved in this section, and item 5 is proved in Proposition 15.9.

Theorem 15.1. *1. For $E = E_j$,*

$$\mathcal{LF}_\beta(\lambda(\gamma_h^E) \cdot h) = \mathcal{LF}_\beta(\lambda(\bar{\gamma}_h^E) \cdot h).$$

As a consequence, write $\lambda_h^E = \lambda(\gamma_h^E)$, then the set $\mathcal{LF}_\beta(\lambda_h^E h)$ is well defined (the definition is independent of the choice of γ_h^E).
2. For each $E > 0$, there exist $-\infty \leq a_E^-(h) \leq a_E^+(h) \leq \infty$ such that

$$\mathcal{LF}_\beta(\lambda_h^E h) = l_E(h)h^* + [a_E^-(h), a_E^+(h)] \ h^\perp.$$

Moreover, the set function $[a_E^-, a_E^+]$ is upper semi-continuous in E.
3. For each $c \in \mathcal{LF}_\beta(\lambda_h^E h)$, $E \neq E_j$, there is a unique c-minimal measure supported on γ_h^E.
4. For each $c \in \mathcal{LF}_\beta(\lambda_h^{E_j} h)$, there are two c-minimal measures supported on $\gamma_h^{E_j}$ and c.
5. For $E > 0$, assume that the torus $\mathbb{T}^2$ is not completely foliated by shortest closed g_E-geodesics in the homology h, then $a_E^+(h) - a_E^-(h) > 0$ and the channel has non-zero width.

We have the following well-known lemma, see for example [47].

Lemma 15.2. *Let $h, h_1, h_2 \in H_1(\mathbb{T}^2, \mathbb{Z})$, then*

- *(Positive homogeneous) $l_E(nh) = hl_E(h)$ for all $n \in \mathbb{N}$.*
- *(Sublinear) $l_E(h_1 + h_2) \leq l_E(h_1) + l_E(h_2)$.*

Assume that γ is a geodesic parametrized according to the Maupertuis principle. First, we note the following useful relation.

$$L(\gamma, \dot{\gamma}) + E = 2(E + U(\gamma)) = \sqrt{g_E(\gamma, \dot{\gamma})}, \tag{15.1}$$

where L denote the associated Lagrangian.

According to [21], the minimal measures for L is in one-to-one correspondence with the minimal measures of $g_E(\varphi, v)$ (considered as a Lagrangian function). On the other hand, any minimal measure $\frac{1}{2}g_E$ with a rational rotation number is supported on a closed geodesic. The following lemma characterizes minimal measures supported on a closed geodesic.

Lemma 15.3. *1. Assume that $c \in H^1(\mathbb{T}^2, \mathbb{R})$ is such that $\alpha_H(c) = E > 0$. Then for any $h \in H_1(\mathbb{T}^2, \mathbb{Z})$,*

$$l_E(h) - \langle c, h \rangle \geq 0.$$

2. *Let γ be a closed geodesic of g_E, $E > 0$, with $[\gamma] = h \in H_1(\mathbb{T}^2, \mathbb{Z})$. Let μ be the probability measure uniformly distributed on the periodic orbit associated to γ. Then given $c \in H^1(\mathbb{T}^2, \mathbb{R})$ with $\alpha_{H^s}(c) = E$,*

$$\mu \text{ is } c - \text{minimal if and only if } \quad l_E(h) - \langle c, h \rangle = 0. \tag{15.2}$$

3. *Let γ be a closed geodesic g_E, $E \geq 0$, with $[\gamma] = h \in H_1(\mathbb{T}^2, \mathbb{Z})$ and $\alpha_{H^s}(c) = E$. Then $\gamma \subset \mathcal{A}_{H^s}(c)$ if and only if (15.2) holds.*

Proof. Let γ be a closed geodesic of g_E, $E > 0$, with $[\gamma] = h$. Assume that with the Maupertuis parametrization, the periodic of γ is T. Let μ be the associated invariant measure, then $\rho(\mu) = h/T$. Assume that $\alpha_{H^s}(c) = E$, by definition, we have

$$\int L d\mu + E \geq \beta(h/T) + \alpha_{H^s}(c) \geq \langle c, h/T \rangle.$$

By (15.1), we have

$$\int L d\mu + E = \frac{1}{T} \int_0^T (L + E)(d\gamma) = \frac{1}{T} \int_0^T \sqrt{g_E(d\gamma)} = l_E(\gamma)/T.$$

Combining the two expressions, we have $l_E(\gamma) - \langle c, h \rangle \geq 0$. By choosing γ such that $l_E(\gamma) = l_E(h)$, statement 1 follows.

To prove statement 2, notice that if μ is c-minimal, then $\alpha_{H^s}(c) = E$ and the equality

$$\int L d\mu + E = \langle c, h/T \rangle$$

holds. Equality (15.2) follows from the same calculation as statement 1.

For $E > 0$, $\gamma \subset \mathcal{A}_{H^s}(c)$ if an only if γ is a minimal measure. Hence we only need to prove statement 3 for $E = 0$. In this case, γ can be parametrized as a homoclinic orbit. $\gamma \subset \mathcal{A}_{H^s}(c)$ if and only if

$$\int_{-\infty}^{\infty} (L - c \cdot v + \alpha(c))(d\gamma) = 0.$$

Since

$$\int_{-\infty}^{\infty} (L - c \cdot v + \alpha(c))(d\gamma) = \int_{-\infty}^{\infty} (L + E)(d\gamma) - \langle c, h \rangle = l_E(h) - \langle c, h \rangle,$$

the statement follows. $\qquad\square$

Proof of Theorem 15.1, item 1–4. By Lemma 15.3, if there are two shortest geodesics γ_h^E and $\bar{\gamma}_h^E$ for g_E, for any c, the invariant measure supported on γ_h^E is c-minimal if and only if the measure on $\bar{\gamma}_h^E$ is c-minimal. This implies statement 1.

Statement 2 follows from the fact that $\mathcal{LF}_\beta(\lambda_h^E h)$ is a closed convex set, and

(15.2).

Statements 3 and 4 follow directly from Lemma 15.3. Item 5 is proved in Proposition 15.9. $\qquad\square$

We also record the following consequence of Theorem 4.1.

Lemma 15.4. *The period* $T_E := T(\gamma_h^E)$ *and length* l_h^E *as functions of* E *are strictly monotone for* $E > 0$.

Proof. Let γ_h^E be a minimal geodesic of g_E, then for any $E' < E$ we have

$$l_{E'}(h) \leq \int \sqrt{g_{E'}(\dot\gamma_h^E)} < \int \sqrt{g_E(\dot\gamma_g^E)} = l_E(h),$$

therefore $l_E(h)$ is strictly monotone.

To prove strict monotonicity of T_E, we apply Theorem 15.1, item 2 to get

$$\mathcal{LF}_\beta(h/T_E) \cdot h = l_h^E h^* \cdot h,$$

where $h^* \cdot h > 0$. Since $\mathcal{LF}_\beta(h/T_E)$ are distinct for different E, $T_E \neq T_{E'}$ for $E \neq E'$, we obtain strict monotonicity. $\qquad\square$

15.2 CHARACTERIZATION OF THE CHANNEL AND THE AUBRY SETS

In this section we provide a precise characterization of the set

$$\mathcal{LF}_\beta(\lambda_h^E h) = l_E(h) \, h^* + [a_E^-(h), a_E^+(h)] \, h^\perp.$$

For each $E > 0$, define

$$d_E^\pm(h) = \pm \inf_{n\to\infty} (l_E(nh \pm \bar{h}) - l_E(nh)).$$

Note that $l_E(nh \pm \bar{h}) - l_E(nh) \leq l_E((n-1)h \pm \bar{h}) + l_E((n-1)h)$ from the sub-additivity and positive homogeneity of l_E; therefore the infimum coincides with the limit. We will omit dependence on h when it is not important.

Lemma 15.5. *The function* $d_E^\pm(h)$ *is continuous in* $E > 0$.

Proof. From the sub-linearity of $l_E(h)$, we get

$$-l_E(\bar{h}) \leq l_E(nh \pm \bar{h}) - l_E(nh) \leq l_E(\bar{h}),$$

as a result, the family $l_E(nh \pm \bar{h}) - l_E(nh)$ as a function of $E \in [E_1, E_2] \subset (0, \infty)$

is equi-continuous and equi-bounded. Then there exists a subsequence such that

$$\lim_{n\to\infty} l_E(nh \pm \bar{h}) - l_E(nh) = d_E^+(h)$$

uniformly. This implies $d_E^+(h)$ is continuous. The same argument works for $d_E^-(h)$. $\square$

Proposition 15.6. *For each $E > 0$, we have*

$$d_E^\pm(h) = a_E^\pm(h).$$

Proof. We first show

$$d_E^-(h) \le a_E^-(h) \le a_E^+(h) \le d_E^+(h).$$

Denoting $c^+ = l_E(h)h^* + a_E^+ h^\perp$, by definition, $l_E(h) - \langle c^+, h \rangle = 0$. By Lemma 15.3, statement 1, for $n \in \mathbb{N}$,

$$0 \le l_E(nh + \bar{h}) - \langle c^+, nh + \bar{h} \rangle = l_E(nh + \bar{h}) - n l_E(h) - \langle c^+, \bar{h} \rangle$$
$$= l_E(nh + \bar{h}) - n l_E(h) - a_E^+.$$

Taking infimum in n, we have $d_E^+ - a_E^+ \ge 0$. Performing the same calculation with $nh + \bar{h}$ replaced by $nh - \bar{h}$, we obtain $0 \le l_E(nh - \bar{h}) - n l_E(h) + a_E^-$, and hence $a_E^- - d_E^- \ge 0$.

We now prove the opposite direction. Taking any $c \in l_E(h)h^* + [d_E^-, d_E^+]h^\perp$, we first show that $\alpha(c) = E$.

Take $\rho \in \mathbb{Q}h + \mathbb{Q}\bar{h}$, then any invariant measure μ with rotation number ρ is supported on some $[\gamma] = m_1 h + m_2 \bar{h}$ with $m_1, m_2 \in \mathbb{Z}$. Let T denote the period. By Lemma 15.7 below,

$$\beta(\rho) + E = l_E(m_1 h + m_2 \bar{h})/T \ge \langle c, m_1 h + m_2 \bar{h} \rangle /T = \langle c, \rho \rangle.$$

Since β is continuous, we have $\alpha(c) = \sup\langle c, \rho \rangle - \beta(\rho) \le E$, where the supremum is taken over all rational ρ's. Since the equality is achieved at $\rho = h$, we conclude that $\alpha(c) = E$.

By Lemma 15.3, statement 2, the measure supported on γ_h^E is c-minimal, and hence $c \in \mathcal{LF}_\beta(\lambda_h^E h)$. $\square$

Recall $h, \bar{h}$ form a basis in $H_1(T^2, \mathbb{Z})$ and the dual of $\bar{h}$ is perpendicular to h and denoted by $h^\perp$.

Lemma 15.7. *For any $c \in l_E(h) h^* + [d_E^-, d_E^+] h^\perp$ and $m_1, m_2 \in \mathbb{Z}$, we have*

$$l_E(m_1 h + m_2 \bar{h}) - \langle c, m_1 h + m_2 \bar{h} \rangle \ge 0.$$

Moreover, if $c \in l_E(h)h^* + (d_E^-, d_E^+)h^\perp$ and $m_1, m_2 \ne 0$, there exists $a > 0$ such

that
$$l_E(m_1 h + m_2 \bar{h}) - \langle c, m_1 h + m_2 \bar{h} \rangle > a > 0.$$

Proof. The inequality for $m_1 = 0$ or $m_2 = 0$ follows from positive homogeneity of l_E. We now assume m_1, m_0.

If $m_2 > 0$, for a sufficiently large $n \in \mathbb{N}$, we have

$$
\begin{aligned}
& l_E(m_1 h + m_2 \bar{h}) - \langle c, m_1 h + m_2 \bar{h} \rangle \\
&= l_E(m_1 h + m_2 \bar{h}) + l_E((nm_2 - m_1)h) - \langle c, nm_2 h + m_2 \bar{h} \rangle \\
&\geq l_E(m_2(nh + \bar{h})) - \langle c, m_2(nh + \bar{h}) \rangle = m_2(l_E(nh + \bar{h}) - \langle c, nh + \bar{h} \rangle) \\
&\geq l_E(nh + \bar{h}) - \langle c, nh + \bar{h} \rangle.
\end{aligned}
$$

Since
$$l_E(nh + \bar{h}) - \langle c, nh + \bar{h} \rangle = l_E(nh + \bar{h}) - n l_E(h) - \langle c, \bar{h} \rangle,$$
for $c \in l_E(h)h^* + (d_E^-, d_E^+)h^\perp$, then there exists $a > 0$ such that

$$\lim_{n \to \infty} l_E(nh + \bar{h}) - n l_E(h) - \langle c, \bar{h} \rangle > a.$$

For $m_2 < 0$, we replace the term $(nm_2 - m_1)h$ with $(-nm_2 - m_1)h$ in the above calculation. $\qquad\square$

The next statement follows from a more general result of Massart (Theorem 1.3 of [57]) for Lagrangian systems on the surface. Here we give a proof in the simpler $\mathbb{T}^2$ case.

Proposition 15.8. *For any $E > 0$ and $c \in l_E(h)h^* + (d_E^-, d_E^+)h^\perp$, we have*

$$\mathcal{A}_{H^s}(c) = \gamma_h^E$$

if E is not a bifurcation value and

$$\mathcal{A}_{H^s}(c) = \gamma_h^E \cup \bar{\gamma}_h^E$$

if E is a bifurcation value.

Proof. We first consider the case when E is not a bifurcation value. Since γ_h^E is the unique closed shortest geodesic, if $\mathcal{A}_{H^s}(c) \supsetneq \gamma_h^E$, it must contain an infinite orbit γ^+. Moreover, as γ_h^E supports the unique minimal measure, the orbit γ^+ must be biasymptotic to γ_h^E. As a consequence, there exist $T_n, T_n' \to \infty$ such that $\gamma^+(-T_n) - \gamma^+(T_n') \to 0$. By closing this orbit using a geodesic, we obtain a closed piece-wise geodesic curve γ_n. Moreover, since γ^+ has no self-intersection, we can arrange it such that γ_n also have no self-intersection. We have

$$\int (L - c \cdot v + \alpha(c))(d\gamma_n) = \int (L + E)(d\gamma_n) - \langle c, [\gamma_n] \rangle = l_E(\gamma_n) - \langle c, [\gamma_n] \rangle.$$

By the definition of the Aubry set, and taking the limit as $n \to \infty$, we have

$$\lim_{n \to \infty} l_E(\gamma_n) - \langle c, [\gamma_n] \rangle = 0.$$

Since γ_n has no self-intersection, we have $[\gamma_n]$ is irreducible. However, this contradicts the strict inequality obtained in Lemma 15.7.

We now consider the case when E is a bifurcation value, and there are two shortest geodesics γ_h^E and $\bar{\gamma}_h^E$. Assume by contradiction that $\mathcal{A}_{H^s}(c) \supseteq \gamma_h^E \cup \bar{\gamma}_h^E$. For mechanical systems on $\mathbb{T}^2$, the Aubry set satisfies an ordering property. As a consequence, there must exist two infinite orbits γ_1^+ and γ_2^+ contained in the Aubry set, where γ_1^+ is forward asymptotic to γ_h^E and backward asymptotic to $\bar{\gamma}_h^E$, and γ_2^+ is forward asymptotic to $\bar{\gamma}_h^E$ and backward asymptotic to γ_h^E. Then there exist $T_n, T_n', S_n, S_n' \to \infty$ such that

$$\gamma_1^+(T_n') - \gamma_2^+(-S_n), \gamma_2^+(S_n') - \gamma_1^+(-T_n) \to 0$$

as $n \to \infty$. The curves $\gamma_{1,2}^+, \gamma_h^E, \bar{\gamma}_h^E$ are all disjoint on $\mathbb{T}^2$. Similar to the previous case, we can construct a piecewise geodesic, non-self-intersecting closed curve γ_n with

$$\lim_{n \to \infty} \int (L - c \cdot v + \alpha(c))(d\gamma_n) = 0.$$

This, however, leads to a contradiction for the same reason as the first case. $\square$

15.3   THE WIDTH OF THE CHANNEL

We show that under our assumptions, the "width" of the channel

$$d_E^+(h) - d_E^-(h) = \inf_{n \in \mathbb{N}} (l_E(nh + \bar{h}) - l_E(nh)) + \inf_{n \in \mathbb{N}} (l_E(nh - \bar{h}) - l_E(nh))$$

is non-zero.

The following statement is a small modification of a theorem of Mather (see [64] and Theorem 1 of [56], for a more general result), we provide a proof using our language.

Proposition 15.9. *For $E > 0$, assume that the torus $\mathbb{T}^2$ is not completely foliated by shortest closed g_E-geodesics in the homology h. Then*

$$d_E^+(h) - d_E^-(h) > 0.$$

Remark 15.10. This is the last item of Theorem 15.1.

Proof. Let $\mathcal{M}$ denote the union of all shortest closed g_E-geodesics in the homology h. We will show that $\mathcal{M} \neq \mathbb{T}^2$ implies $d_E^+(h) - d_E^-(h) > 0$. Omit h

dependence. For $n \in \mathbb{N}$, denote

$$d_n = (l_E(nh + \bar{h}) - l_E(nh)) + (l_E(nh - \bar{h}) - l_E(nh)).$$

Assume by contradiction that $\inf d_n = \lim d_n > 0$.

Let γ_0 be a shortest geodesic in homology h. We denote $\tilde{\gamma}_0$ its lift to the universal cover, and use "$\leq_{\tilde{\gamma}}$" to denote the order on $\tilde{\gamma}$ defined by the flow. Let γ_1 and γ_2 be shortest curves in the homology $nh + \bar{h}$ and $nh - \bar{h}$ respectively, and let T_1 and T_2 be their periods. γ_i depends on n but we will not write it down explicitly.

Let $\tilde{\gamma}_i$, $i = 0, 1, 2$ denote a lift of γ_i to the universal cover. Using the standard curve shortening lemma in Riemmanian geometry, it's easy to see that $\tilde{\gamma}_i$ and $\tilde{\gamma}_j$ may intersect at most once. Let $a \in \gamma_0 \cap \gamma_1$ and lift it to the universal cover without changing its name. Let $b \in \gamma_0 \cap \gamma_2$, and we choose a lift in $\tilde{\gamma}_0$ by the largest element such that $b \leq a$. We now choose the lifts $\tilde{\gamma}_i$ of γ_i, $i = 1, 2$, by the relations $\tilde{\gamma}_1(0) = a$ and $\tilde{\gamma}_2(T_2) = b$.

We have for $1 \leq k \leq 2n$, $\tilde{\gamma}_2(T_2) + kh > \tilde{\gamma}_1(0)$ and

$$\tilde{\gamma}_2(0) + kh = b - (nh - \bar{h}) + kh \leq_{\tilde{\gamma}_0} a + nh + \bar{h} = \tilde{\gamma}_1(T_1).$$

As a consequence, $\tilde{\gamma}_2 + kh$ and $\tilde{\gamma}_1$ has a unique intersection. Let

$$x_k = (\tilde{\gamma}_2 + kh) \cap \tilde{\gamma}_1, \quad \bar{x}_k = (\tilde{\gamma}_2 + kh) \cap (\tilde{\gamma}_1 - h).$$

We have x_k is in increasing order on $\tilde{\gamma}_1$ and $\bar{x}_k$ is in decreasing order after projection to γ_2 (see Figure 15.1) . Define

$$\gamma_k^* = (\tilde{\gamma}_2 + kh)|[\bar{x}_k, x_k] * \tilde{\gamma}_1|[x_k, x_{k+1}],$$

and let γ_k^* be its projection to $\mathbb{T}^2$. We have $[\gamma_k] = h$ and

$$\sum_{k=1}^{2n} l_E(\gamma_k^*) = l_E(\gamma_1) + l_E(\gamma_2).$$

Using $l_E(\gamma_k) \geq l_E(h)$ and $l_E(\gamma_1) + l_E(\gamma_2) \leq 2n l_E(h) + d_n$, we obtain

$$l_E(h) \leq l_E(\gamma_k) \leq l_E(h) + d_n.$$

Any connected component in the completement of $\mathcal{M}$ is diffeomorphic to an annulus. Pick one such annulus, and let $b > 0$ denote the distance between its boundaries. Since γ_1 intersects each boundary once, there exists a point $y_n \in \gamma_1$ such that $d(y_n, \mathcal{M}) = b/2$. Since $\gamma_1 \subset \bigcup_k \gamma_k^*$ there exists some γ_k^* containing y_n. By taking a subsequence if necessary, we may assume $y_n \to y_* \notin \mathcal{M}$. Using the above discussion, we have

$$l_E(h) \leq \inf_{y_* \in \gamma, [\gamma] = h} l_E(\gamma) \leq \inf_n l_E(h) + d_n = l_E(h).$$

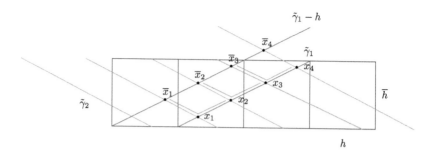

Figure 15.1: Proof of Proposition 15.9; the L-shaped curves are $\tilde{\gamma}_k^*$'s.

Since the collection of rectifiable curves satisfying $y_* \in \gamma, [\gamma] = h$ is compact, and that l_E is lower semi-continuous over such curves, there exists a rectifiable curve γ_* containing y_* with $l_E(\gamma_*) = l_E(h)$, hence γ_* is a shortest curve. But $y_* \notin \mathcal{M}$, leading to a contradiction. □

Proposition 15.9 clearly applies to the slow system, as there are either one or two shortest geodesics.

15.4 THE CASE $E = 0$

We now extend the earlier discussions to the case $E = 0$. While the function $a_E^\pm$ is not defined at $E = 0$, the function $d_E^\pm$ is well defined at $E = 0$. Recall $h, \bar{h}$ form a basis in $H_1(T^2, \mathbb{Z})$ and the dual of $\bar{h}$ is perpendicular to h and denoted by $h^\perp$.

Proposition 15.11. *The properties of the channel and the Aubry sets depend on the type of homology h.*

1. *Assume h is simple and critical.*

   a) $d_0^+(h) - d_0^-(h) > 0$.
   b) $l_0(h)h^* + [d_0^-(h), d_0^+(h)] h^\perp \subset \mathcal{LF}_\beta(0)$.
   c) *For $c \in l_0(h) h^* + [d_0^-(h), d_0^+(h)] h^\perp$, we have $\gamma_h^0 \subset \mathcal{A}_{H^s}(c)$;*
      For $c \in l_0(h)h^* + (d_0^-(h), d_0^+(h)) h^\perp$, we have $\gamma_h^0 = \mathcal{A}_{H^s}(c)$.

2. *Assume h is simple and non-critical.*

   a) $d_0^+(h) - d_0^-(h) > 0$.
   b) $l_0(h)h^* + [d_0^-(h), d_0^+(h)] h^\perp \subset \mathcal{LF}_\beta(0)$.
   c) *For $c \in l_0(h)h^* + [d_0^-(h), d_0^+(h)] h^\perp$, we have $\gamma_h^0 \cup \{0\} \subset \mathcal{A}_{H^s}(c)$;*
      For $c \in l_0(h)h^* + (d_0^-(h), d_0^+(h))h^\perp$, we have $\gamma_h^0 \cup \{0\} = \mathcal{A}_{H^s}(c)$.
   d) *The function $d_E^\pm(h)$ is right-continuous at $E = 0$.*

3. *Assume h is non-simple and $h = n_1 h_1 + n_2 h_2$, with h_1, h_2 simple.*

  a) $d_0^+(h) = d_0^-(h)$. *Moreover, let $c^*(h) = l_E(h_1)h_1^* + l_E(h_2)h_2^*$, where (h_1^*, h_2^*) is the dual basis to (h_1, h_2), then*

$$c^* = l_E(h)h^* + d_0^\pm(h)h^\perp,$$

  where $h^\perp$ is a unit vector perpendicular to h.
  b) $\gamma_{h_1}^0 \cup \gamma_{h_2}^0 = \mathcal{A}_{H^s}(c^*)$.
  c) $d_0^+(h_1) - d_0^-(h_1) > 0$ *with*

$$l_E(h_1)h_1^* + d_0^+(h_1)h_2^* = c^*.$$

Before proving Proposition 15.11, we first explain how the proof of Proposition 15.9 can be adapted to work even for $E = 0$.

Lemma 15.12. *Assume that there is a unique g_0−shortest geodesic in the homology h. Then*

$$d_0^+(h) - d_0^-(h) > 0.$$

Proof. We will try to adapt the proof of Proposition 15.9. Let γ_0, γ_1, and γ_2 be shortest geodesics in homologies h, $nh + \bar{h}$ and $nh - \bar{h}$, respectively. We choose an arbitrary parametrization for γ_i on $[0, T]$. Note that the parametrization is just continuous in general.

The proof of Proposition 15.9 relies only on the property that lifted shortest geodesics intersect at most once. For $E = 0$, we will rely on a weaker property.

Let $\tilde{\gamma}_i$ be the lifts to the universal cover $\mathbb{R}^2$. The degenerate point $\{0\}$ lifts to the integer lattice $\mathbb{Z}^2$. Since g_0 is a Riemannian metric away from the integers, using the shortening argument, we have: If γ_i intersect γ_j at more than one point, then either the intersections occur only at integer points, or the two curve coincide on a segment with integer end points.

Let $a_0 \in \gamma_0 \cap \gamma_1$ and let $\tilde{\gamma}_0$ and γ_1 be lifts with $\tilde{\gamma}_0(0) = \tilde{\gamma}_1(0) = a_0$. If $a \notin \mathbb{Z}^2$, then it is the only intersection between the two curves. If $a_0 \in \mathbb{Z}^2$, we define a_0' to be the largest intersection between $\tilde{\gamma}_0|[0, T)$ and $\tilde{\gamma}_1$ according to the order on $\tilde{\gamma}_0$. a_0' is necessarily an integer point, and since $a_0' \in \tilde{\gamma}_0$, there exists $n_1 < n$ such that $a_0' - a_0 = n_0 h$. Moreover, using the fact that $\tilde{\gamma}_0$ is minimizing, we have

$$l_0(\tilde{\gamma}_0|[a_0, a_0']) = l_0(\tilde{\gamma}_1|[a_0, a_0']).$$

We now apply a similar argument to $\tilde{\gamma}_0 + \bar{h}$ and $\tilde{\gamma}_1$. Let $a_1 = \tilde{\gamma}_0(T) + \bar{h} = \tilde{\gamma}_1(T)$ and let a_1' be the smallest intersection between $\tilde{\gamma}_0|(0, T]$ and $\tilde{\gamma}_1$. Then there exists $n_1 \in \mathbb{N}$, $n_0 + n_1 < n$, such that $a_1 - a_1' = n_1 h$. Moreover,

$$l_0((\tilde{\gamma}_0 + \bar{h})|[a_1', a_1]) = l_0(\tilde{\gamma}_1|[a_1', a_1]).$$

Let $\tilde{\eta}_1 = \tilde{\gamma}_1|[a_0', a_1']$ and η_1 be its projection. We have $[\eta] = (n - n_0 - n_1)h + \bar{h} =:$

$m_1 h + \bar{h}$, and
$$l_0(\eta_1) - m_1 l_0(h) = l_0(\gamma_1) - n l_0(h).$$

The curve η_1 has the property that it intersects γ_0 only once. Applying the same argument to γ_2, we obtain a curve η_2 with $[\eta_2] = m_2 h - \bar{h}$, and

$$l_0(\eta_2) - m_2 l_0(h) = l_0(\gamma_2) - n l_0(h).$$

To proceed as in the proof of Proposition 15.9, we show that if $\tilde{\eta}_1$ and $\tilde{\eta}_2$ are lifts of η_1 and η_2 with the property that

$$\tilde{\eta}_1(0), \tilde{\eta}_2(T) \in \{\tilde{\gamma}_0(t)\}, \quad \tilde{\eta}_1(T), \tilde{\eta}_2(0) \in \{\tilde{\gamma}_0(t) + \bar{h}\},$$

then $\tilde{\eta}_1$ and $\tilde{\eta}_2$ intersect only once. Indeed, there are no integer points between $\tilde{\gamma}_0$ and $\tilde{\gamma}_0 + \bar{h}$.

We have

$$l_0(\eta_1) - m_1 l_0(h) + l_0(\eta_2) - m_2 l_0(h) = d_n,$$

where d_n is as defined in Proposition 15.9. Assuming $\inf d_n = 0$, we proceed as in the proof of Proposition 15.9 and obtain curves $[\gamma_k] = h$, positive distance away from γ_0, such that

$$l_0(h) \le l_0(\gamma_k) \le l_0(h) + d_n.$$

This leads to a contradiction.   □

Proof of Proposition 15.11. Case 1, h is simple and critical.
  (a) This follows from Lemma 15.12.
  (b) We note that Lemma 15.7 depends only on positive homogeinity and sub-additivity of $l_E(h)$, and hence applies even when $E = 0$. We obtain for $c \in l_0(h)h^* + [d_0^-(h), d_0^+(h)]\bar{h}^*$

$$l_0(h') - \langle c, h' \rangle \ge 0, \forall h' \in H_1(\mathbb{T}^2, \mathbb{Z}^2).$$

Since $l_E(h)$ is strictly increasing, we obtain $l_E(h') - \langle c, h' \rangle > 0$ for $E > 0$. By Lemma 15.3, there are no c-minimal measures with energy $E > 0$. As a consequence, $\alpha(c) = 0$. Since $\{0\}$ is a c-minimal measure with rotation number 0, we conclude $l_0(h)h^* + [d_0^-(h), d_0^+(h)]\bar{h}^* \subset \mathcal{LF}_\beta(0)$.
  (c) Since we proved $\alpha(c) = 0$, the first conclusion follows from Lemma 15.3. For the second conclusion, we verify that the proof of Proposition 15.8 for non-bifurcation values applies to this case.
  (d) The set function $[d_E^-(h), d_E^+(h)]$ is upper semi-continuous at $E = 0$ from the right, by definition. We will show that it is continuous. Assume by contradiction that

$$[\liminf_{E \to 0+} d_E^-(h), \limsup_{E \to 0+} d_E^+(h)] \subsetneq [d_0^-(h), d_0^+(h)].$$

Then there exists $c \in l_0(h)h^* + (d_0^-(h), d_0^+(h))\bar{h}^*$ and

$$c(E) \notin l_E(h)h^* + [d_E^-(h), d_E^+(h)]h^\perp$$

such that $c(E) \to c$. By part (c), the Aubry set $\mathcal{A}_{H^s}(c)$ supports a unique minimal measure. By Proposition 7.3, the Aubry set is upper semi-continuous in c. Hence any limit point of $\mathcal{A}_{H^s}(c(E))$ as $E \to 0$ is in $\mathcal{A}_{H^s}(c)$. This implies that $\tilde{\mathcal{A}}_{H^s}(c(E))$ approaches γ_h^E as $E \to 0$. Since γ_h^E is the unique closed geodesic in a neighborhood of itself, we conclude that $\tilde{\mathcal{A}}_{H^s}(c(E)) = \gamma_h^E$ for sufficiently small E. But this is a contradiction with $c(E) \notin l_E(h)h^* + [d_E^-(h), d_E^+(h)]h^\perp$.

Case 2, h is simple and non-critical.

(a) This follows from Lemma 15.12.

(b) The proof is identical to that of Case 1.

(c) For the first conclusion, we can directly verify that $\gamma_h^0 \subset \mathcal{A}_{H^s}(c)$ and $\{0\} \subset \mathcal{A}_{H^s}(c)$. For the second conclusion, we note that proof of Proposition 15.8 for bifurcation values applies to this case.

Case 3, h is non-simple with $h = n_1 h_1 + n_2 h_2$.

(a) Assume that $\bar{h} = m_1 h_1 + m_2 h_2$ for some $m_1, m_2 \in \mathbb{Z}$. For sufficiently large $n \in \mathbb{N}$, we have $nh \pm \bar{h} \in \mathbb{N}h_1 + \mathbb{N}h_2$. As a consequence,

$$\begin{aligned}
l_0(nh \pm \bar{h}) &- l_0(nh) \\
&= (nn_1 \pm m_1)l_0(h_1) + (nn_2 \pm m_2)l_0(h_2) - (nn_1 l_0(h_1) + nn_2 l_0(h_2)) \\
&= \pm m_1 l_0(h_1) \pm m_2 l_0(h_2).
\end{aligned}$$

We obtain $d_0^+(h) - d_0^-(h) = 0$ by definition.

We check directly that

$$l_0(h) - \langle c^*, h \rangle = 0.$$

Since $l_0(h)h^* + d_0^-(h)h^\perp = l_0(h)h^* + d_0^+(h)h^\perp$ is the unique c with this property, the second claim follows.

(b) We note that any connected component of the complement to $\gamma_{h_1}^0 \cup \gamma_{h_2}^0$ is contractible. If $\mathcal{A}_{H^s}(c)$ has other components, the only possibility is a contractible orbit bi-asymptotic to $\{0\}$. However, such an orbit can never be minimal, as the fixed point $\{0\}$ has smaller action.

(c) The statement $d_0^+(h_1) - d_0^-(h_1) > 0$ follows from part 1(a). for the second claim, we compute

$$d_0^+(h_1) = \inf_n l_0(nh_1 + h_2) - l_0(nh_2) = l_0(h_2)$$

and the claim follows. $\qquad\square$

Appendix

Notations

We provide a list of notations for the reader's convenience.

FORMULATION OF THE MAIN RESULT

(θ, p, t)	A point in the phase space $\mathbb{T}^n \times \mathbb{R}^n \times \mathbb{T}_\varpi$.
ϕ_H^t	The flow of $H(\theta, p, t)$ on $\mathbb{T}^n \times \mathbb{R}^n \times \mathbb{T}_\varpi$, or the flow of $H(\theta, p)$ on $\mathbb{T}^n \times \mathbb{R}^n$.
$\phi_H^{s,t}$	The Hamiltonian diffeomorphism of $H(\theta, p, t)$ from time s to times t.
ϕ_H	The time-ϖ map of ϕ_H^t. Equal to the Poincaré map of ϕ_H^t on $\{t = 0\}$.
$H_\epsilon = H_0 + \epsilon H_1$	Nearly integrable system.
D	A fixed constant controlling the norm and convexity of H_0.
$\mathbb{Z}_*^3 = \mathbb{Z}^3 \setminus \{(0,0,1)\}\mathbb{Z}$	Set of integer vectors defining a resonance relation.
$k = (k^1, k^2, k^0)$	Resonance vectors, contained in $\mathbb{Z}_*^3$.
S_k	Single-resonance surface in the action space given by $k \in \mathbb{Z}_*^3$.
Γ_k	Singe-resonance segment, a closed segment contained in S_k.
Γ_{k_1, k_2}	Double resonance point in the action space given by $k_1, k_2 \in \mathbb{Z}_*^3$.
$\mathcal{S}^r$	The unit sphere of the C^r functions.
$\mathcal{P}$	Diffusion path consisting of segments of single resonance segments.
$\mathcal{K} = \{(k, \Gamma_k)\}$	Collection of resonances and resonance segments making up a diffusion path.
$\mathcal{U} = \mathcal{U}(\mathcal{P})$	An open and dense set in $\mathcal{S}^r$ defining the "non-degenerate perturbations" relative to a diffusion path.
$\mathcal{U}_{SR}^\lambda(k_1, \Gamma_{k_1})$	Set of $H_1 \in \mathcal{S}^r$ satisfying the quantitative non-resonance conditions $[SR1_\lambda]$–$[SR3_\lambda]$ relative to the resonant segment (k_1, Γ_{k_1}).
$\mathcal{U}_{DR}(k_1, k_2)$	Set of $H_1 \in \mathcal{S}^r$ satisfying the non-degeneracy conditions $[DR1^h] - [DR3^h]$ and $[DR1^c]$–$[DR4^c]$.

$\mathcal{V} = \mathcal{V}(\mathcal{U}, \epsilon_0)$	A "cusp" set of perturbations, equal to $\{\epsilon H_1 : H_1 \in \mathcal{U}, 0 < \epsilon < \epsilon_0(H_1)\}$.
$\mathcal{K}^{\text{st}}(k_1, \Gamma_{k_1}, \lambda)$	Set of strong additional resonances relative to the resonance segment Γ_{k_1}, for a perturbation $H_1 \in \mathcal{U}_{SR}^{\lambda}(k_1, \Gamma_{k_1})$.
B_σ	Ball in Euclidean space $\mathbb{R}^n$.
$\mathcal{V}_\sigma^r, \mathcal{V}_\sigma$	Ball in the functional space C^r. When superscript is not indicated, then stands for C^r where r is from the main theorem.

WEAK KAM AND MATHER THEORY

$\varpi = \varpi(H)$	The period of a time periodic Hamiltonian, i.e. $H(\theta, p, t + \varpi) = H(\theta, p, t)$.
$\mathbb{T}_\varpi = \mathbb{R}/(\varpi\mathbb{Z})$	Torus with period ϖ.
$\mathbb{H} = \mathbb{H}(D)$	A family of Hamiltonians satisfying uniform conditions depending on the parameter $D > 1$.
$L = L_H$	The Lagrangian of H.
$L_{H,c}$	The "penalized" Lagrangian $L_H(\theta, v, t) - c \cdot v$.
$A_{H,c}(x, s, y, t)$	The minimal action for the Lagrangian $L_{H,c}$.
$\alpha_H(c), \alpha_L(c)$	Mather's alpha function.
$\beta_H(\rho), \beta_L(\rho)$	Mather's beta function.
$h_{H,c}(x, s, y, t)$	The time-dependent Peierl's barrier function.
$h_{H,c}(x, y)$	The discrete time Peierl's barrier function, equal to $h_{H,c}(x, 0, y, 0)$.
$d_{H,c}(x, s, y, t)$	Mather's semi-distance
$T_c^{s,t}u(x)$	Lax-Oleinik semi-group defined on $\mathbb{T}^n$.
$\partial^+ u(x)$	The supergradient of a semi-concave function at x.
$\mathcal{G}_{c,w}, \ w = w(\theta, t)$	The time-dependent pseudograph as a subset of $\mathbb{T}^n \times \mathbb{R}^n \times \mathbb{R}$, or $\mathbb{T}^n \times \mathbb{R}^n \times \mathbb{T}_\varpi$.
$\tilde{\mathcal{I}}(c, w)$	The maximum invariant set contained in the psudograph $\mathcal{G}_{c,w}$.
$\widetilde{\mathcal{M}}_H(c), \tilde{\mathcal{A}}_H(c), \tilde{\mathcal{N}}_H(c)$	The continuous (Hamiltonian) Mather, Aubry, and Mañe set, defined on $\mathbb{T}^n \times \mathbb{R}^n \times \mathbb{T}_{\varpi(H)}$.
$\widetilde{\mathcal{M}}_H^0(c), \tilde{\mathcal{A}}_H^0(c), \tilde{\mathcal{N}}_H^0(c)$	The discrete (Hamiltonian) Mather, Aubry, and Mañe set, defined on $\mathbb{T}^n \times \mathbb{R}^n$, invariant under the map $\phi_H = \phi_H^{\varpi(H)}$.
$\tilde{\mathcal{S}}(H, c)$	A static class of the Aubry set $\tilde{\mathcal{A}}_H(c)$.
$c \vdash c'$	The forcing relation.
$c \dashv\vdash c'$	The forcing equivalence relation.

SINGLE RESONANCE

$[H_1]_{k_1}$	The average of H_1 relative to the resonance k_1.
$[H_1]_{k_1,k_2}$	The average of H_1 relative to the double resoannce k_1, k_2.
Φ_ϵ	The averaging coordinate change at single resonance.
N_ϵ	Normal form under the coordinate change. Same notation is used at double resonance.
$Z_{k_1}(\theta^s, p, t)$	The resonant component of H_1 relative to the resonance k_1.
$R(\theta, p, t)$	The remainder in the single resonance normal form.
$\mathcal{K}^{\text{st}}(k_1, \Gamma_{k_1}, K)$	Set of strong additional resonance k_2 intersecting Γ_{k_1} with norm at most K, plus any resonances in the diffusion path that intersect Γ_{k_1}.
$\Gamma_{k_1}^{SR}$	The punctured resonance segment after removing $O(\sqrt{\epsilon})$ neighborhoods of strong double resonances.
$\|\cdot\|_{C_I^r}$	The rescaled C^r norm where the derivatives in the action variable are rescaled by $\sqrt{\epsilon}$.
T_ω	The period of the rational vector $(\omega, 1) \in \mathbb{R}^3$.
$B \in SL(3, \mathbb{Z})$	An integer matrix defining the linear coordinate change relative to a double resonance k_1, k_2.
$(\theta^s, \theta^f, p^s, p^f, t)$	The coordinate at single resonance after taking a linear coordinate change. The resonance becomes $(1, 0, 0) \cdot (\omega, 1) = 0$ in this coordinate.
$(\Theta^s, P^s)(\theta^f, p^f, t)$	Normally hyperbolic invariant cylinder for the single resonant normal form, parametrized using the (θ^f, p^f, t) variables.

DOUBLE RESONANCE

Φ_ϵ	The averaging coordinate change at double resonance.
N_ϵ	The normal form at double resonance.
Φ_L	The linear coordinate change corresponding to the double resonance k_1, k_2.
$K(I)$	The kinetic energy of the slow mechanical system.
$U(\varphi)$	The potential function of the slow mechanical system.
g_E	The Jacobi metric at energy E of the slow mechanical system.
h	A homology class in $H^1(\mathbb{T}^2, \mathbb{Z})$.
γ_h^E	Shortest curves of the Jacobi metric in homology class h.
η_h^E	The Hamiltonian periodic orbit corresponding to the geodesic γ_h^E.

$\Phi_{\text{loc}}^{ij}, i, j \in \{+, -\}$ Local maps near the saddle fixed point of H^s.

Φ_{glob} Global map along a homoclinic orbit η of H^s.

$\bar{c}_h(E)$ Curve of cohomologies chosen in the channel of h.

$\bar{\Gamma}_h$ Choice of cohomologies along the homology h.

$\bar{\Gamma}_h^e$ In the non-simple case, choice of cohomology curve above energy e.

$\bar{\Gamma}_{h_1}^{e,\mu}$ In the non-simple case, choice of cohomology along the adjacent simple homology h_1.

Φ_L^* The relation between cohomology class and alpha function after the coordinate change Φ_L.

$\Phi_{L,H_\epsilon^s}^*$ The relation between the cohomology classes for the coordinate change Φ_L. Depends on the alpha function of the system H_ϵ^s.

Γ_{k_1,k_2}^{DR} The choice of cohomology classes at a double resonance k_1, k_2.

References

[1] Abraham, R., J. E. Marsden, and T. S. Raţiu (1983). *Manifolds, tensor analysis, and applications*, Volume 2 of *Global Analysis Pure and Applied: Series B*. Addison-Wesley Publishing.

[2] Amold, V., V. Kozlov, and A. Neishtadt (1988). Mathematical aspects of classical and celestial mechanics. *Accomplishments in Science and Engineering 3*, 117–122.

[3] Arnaud, M.-C. (2010). Green bundles and related topics. In *Proceedings of the International Congress of Mathematicians*, Volume III, pp. 1653–1679. Hindustan Book Agency, New Delhi.

[4] Arnold, V. I. (1963). Small denominators and problems of stability of motion in classical and celestial mechanics. *Akademiya Nauk SSSR i Moskovskoe Matematicheskoe Obshchestvo. Uspekhi Matematicheskikh Nauk 18*(6 (114)), 91–192.

[5] Arnold, V. I. (1964). Instability of dynamical systems with many degrees of freedom. *Doklady Akademii Nauk SSSR 156*, 9–12.

[6] Arnold, V. I. (1968). The stability problem and ergodic properties for classical dynamical systems. In *Vladimir I. Arnold: Collected Works*, pp. 107–113. Springer-Verlag.

[7] Arnold, V. I. (1994). Mathematical problems in classical physics. In *Trends and perspectives in applied mathematics*, pp. 1–20. Springer-Verlag.

[8] Arnold, V. I. (2013). *Mathematical methods of classical mechanics*, Volume 60. Springer Science+Business Media.

[9] Bernard, P. (1996). Perturbation d'un hamiltonien partiellement hyperbolique. *Comptes Rendus de l'Académie des Sciences. Série I. Mathématique 323*(2), 189–194.

[10] Bernard, P. (2007). Symplectic aspects of Mather theory. *Duke Mathematical Journal 136*(3), 401–420.

[11] Bernard, P. (2008). The dynamics of pseudographs in convex Hamiltonian systems. *Journal of the American Mathematical Society 21*(3), 615–669.

[12] Bernard, P. (2010). On the conley decomposition of Mather sets. *Revista Matematica Iberoamericana 26*(1), 115–132.

[13] Bernard, P., V. Kaloshin, and K. Zhang (2016). Arnold diffusion in arbitrary degrees of freedom and normally hyperbolic invariant cylinders. *Acta Mathematica 217*(1), 1–79.

[14] Berti, M. and P. Bolle (2002). A functional analysis approach to arnold diffusion. In *Annales de l'Institut Henri Poincare (C) Non Linear Analysis*, Volume 19, pp. 395–450. Elsevier.

[15] Bessi, U. (1997). Arnold's diffusion with two resonances. *Journal of Differential Equations 137*(2), 211–239.

[16] Biasco, L., L. Chierchia, and D. Treschev (2006). Stability of nearly integrable, degenerate Hamiltonian systems with two degrees of freedom. *Journal of Nonlinear Science 16*(1), 79–107.

[17] Birkhoff, G. (1950). *Collected Mathematical Papers*, Volume 2 of *Collected Mathematical Papers*. American Mathematical Society.

[18] Bolotin, S. and P. Rabinowitz (1998). A variational construction of chaotic trajectories for a hamiltonian system on a torus. *Bollettino dell'Unione Matematica Italiana 1*(3), 541–570.

[19] Bounemoura, A. (2010). Nekhoroshev estimates for finitely differentiable quasi-convex Hamiltonians. *Journal of Differential Equations 249*(11), 2905–2920.

[20] Bourgain, J. and V. Kaloshin (2005). On diffusion in high-dimensional hamiltonian systems. *Journal of Functional Analysis 229*(1), 1–61.

[21] Carneiro, M. D. (1995). On minimizing measures of the action of autonomous lagrangians. *Nonlinearity 8*(6), 1077.

[22] Castejón, O. and V. Kaloshin (2015). Random iteration of maps on a cylinder and diffusive behavior. *arXiv preprint arXiv:1501.03319*.

[23] Cheng, C.-Q. (2017a). Dynamics around the double resonance. *Cambridge Journal of Mathematics 5*(2), 153–228.

[24] Cheng, C.-Q. (2017b). Uniform hyperbolicity of invariant cylinder. *Journal of Differential Geometry 106*(1), 1–43.

[25] Cheng, C.-Q. (2019). The genericity of Arnold diffusion in nearly integrable Hamiltonian systems. *Asian Journal of Mathematics 23*(3), 401–438.

[26] Cheng, C.-Q. and J. Yan (2004). Existence of diffusion orbits in a priori unstable hamiltonian systems. *Journal of Differential Geometry 67*(3), 457–

517.

[27] Cheng, C.-Q. and J. Yan (2009). Arnold diffusion in Hamiltonian systems: a priori unstable case. *Journal of Differential Geometry 82*(2), 229–277.

[28] Contreras, G. and R. Iturriaga (1999). Convex Hamiltonians without conjugate points. *Ergodic Theory and Dynamical Systems 19*(4), 901–952.

[29] Contreras, G., R. Iturriaga, and H. Sanchez-Morgado (2013). Weak solutions of the Hamilton-Jacobi equation for time periodic lagrangians. *arXiv preprint arXiv:1307.0287*.

[30] Davletshin, M. N. and D. V. Treshchev (2016). Arnold diffusion in a neighborhood of low-order resonances. *Trudy Matematicheskogo Instituta Imeni V. A. Steklova. Rossiĭskaya Akademiya Nauk 295*(Sovremennye Problemy Mekhaniki), 72–106.

[31] Delshams, A., R. de la Llave, and T. M. Seara (2006). A geometric mechanism for diffusion in Hamiltonian systems overcoming the large gap problem: heuristics and rigorous verification on a model. *Memoirs of the American Mathematical Society 179*(844), 1–141.

[32] Delshams, A., M. Gidea, R. de la Llave, and T. M. Seara (2008). Geometric approaches to the problem of instability in hamiltonian systems. an informal presentation. In *Hamiltonian dynamical systems and applications*, pp. 285–336. Springer.

[33] Delshams, A. and G. Huguet (2009). Geography of resonances and arnold diffusion in a priori unstable hamiltonian systems. *Nonlinearity 22*(8), 1997.

[34] Delshams, A. and R. G. Schaefer (2017, Jan). Arnold diffusion for a complete family of perturbations. *Regular and Chaotic Dynamics 22*(1), 78–108.

[35] Delshams, A. and R. G. Schaefer (2018). Arnold diffusion for a complete family of perturbations with two independent harmonics. *Discrete and Continuous Dynamical Systems. Series A 38*(12), 6047–6072.

[36] Ehrenfest, P. and T. Ehrenfest (2002). *The conceptual foundations of the statistical approach in mechanics*. Courier Corporation.

[37] Fathi, A. (2008). *Weak KAM theorem in Lagrangian dynamics, preliminary version number 10*. Book preprint.

[38] Féjoz, J., M. Guàrdia, V. Kaloshin, and P. Roldán (2016). Kirkwood gaps and diffusion along mean motion resonances in the restricted planar three-body problem. *Journal of the European Mathematical Society 18*(10), 2315–2403.

[39] Fermi, E. (1923). Dimostrazione che in generale un sistema meccanico

normale è quasiergodico. *Il Nuovo Cimento (1911-1923) 25*(1), 267–269.

[40] Gallavotti, G. (2001). Fermi and the ergodic problem. In *Proceedings of the International Conferente Enrico Fermi and the Universe of Physics*. ENEA.

[41] Gelfreich, V. and D. Turaev (2008). Unbounded energy growth in Hamiltonian systems with a slowly varying parameter. *Communications in Mathematical Physics 283*(3), 769–794.

[42] Gelfreich, V. and D. Turaev (2017). Arnold diffusion in a priori chaotic symplectic maps. *Communications in Mathematical Physics 353*(2), 507–547.

[43] Gidea, M. and R. De La Llave (2006). Topological methods in the instability problem of Hamiltonian systems. *Discrete and Continuous Dynamical Systems 14*(2), 295.

[44] Gidea, M. and J.-P. Marco (2017). Diffusion along chains of normally hyperbolic cylinders. *arXiv preprint arXiv:1708.08314*.

[45] Guardia, M., V. Kaloshin, and J. Zhang (2016). A second order expansion of the separatrix map for trigonometric perturbations of a priori unstable systems. *Communications in Mathematical Physics 348*(1), 321–361.

[46] Guzzo, M., E. Lega, and C. Froeschl (2009). A numerical study of the topology of normally hyperbolic invariant manifolds supporting arnold diffusion in quasi-integrable systems. *Physica D: Nonlinear Phenomena 238*(17), 1797 – 1807.

[47] Hedlund, G. A. (1932). Geodesics on a two-dimensional riemannian manifold with periodic coefficients. *Annals of Mathematics 33*(4), 719–739.

[48] Hirsch, M. W. (2012). *Differential topology*, Volume 33. Springer Science+Business Media.

[49] Hirsch, M. W., C. C. Pugh, and M. Shub (2006). *Invariant manifolds*, Volume 583. Springer Science+Business Media.

[50] Kaloshin, V., J. N. Mather, E. Valdinoci, et al. (2004). Instability of resonant totally elliptic points of symplectic maps in dimension 4. *Astérisque 297*, 79–116.

[51] Kaloshin, V., J. Zhang, and K. Zhang (2015). Normally hyperbolic invariant laminations and diffusive behaviour for the generalized arnold example away from resonances. *arXiv preprint arXiv:1511.04835*.

[52] Kaloshin, V. and K. Zhang (2018). Dynamics of the dominant Hamiltonian. *Bulletin de la Société Mathématique de France 146*(3), 517–574.

[53] Lochak, P. (1993). Hamiltonian perturbation theory: periodic orbits, reso-

nances and intermittency. *Nonlinearity 6*(6), 885–904.

[54] Marco, J. (2016a). Arnold diffusion for cusp-generic nearly integrable convex systems on $\mathbb{A}^3$. *Preprint available at https://arxiv. org/abs/1602.02403*.

[55] Marco, J.-P. (2016b). Chains of compact cylinders for cusp-generic nearly integrable convex systems on $\mathbb{A}^3$. *arXiv preprint arXiv:1602.02399*.

[56] Massart, D. (2003). On aubry sets and mathers action functional. *Israel Journal of Mathematics 134*(1), 157–171.

[57] Massart, D. (2011). Aubry sets vs. Mather sets in two degrees of freedom. *Calculus of Variations and Partial Differential Equations 42*(3-4), 429–460.

[58] Mather, J. Personal communication.

[59] Mather, J. (1996). Variational construction of trajectories for time periodic Lagrangian systems on the two torus.

[60] Mather, J. N. (1991a). Action minimizing invariant measures for positive definite Lagrangian systems. *Mathematische Zeitschrift 207*(1), 169–207.

[61] Mather, J. N. (1991b). Variational construction of orbits of twist diffeomorphisms. *Journal of the American Mathematical Society 4*(2), 207–263.

[62] Mather, J. N. (2004). Arnold diffusion. I: Announcement of results. *Journal of Mathematical Sciences 124*(5), 5275–5289.

[63] Mather, J. N. (2008). Arnold diffusion. II.

[64] Mather, J. N. (2010). Order structure on action minimizing orbits. In *Symplectic topology and measure preserving dynamical systems*, Volume 512 of *Contemporary Mathematics*, pp. 41–125. American Mathematical Society.

[65] Mather, J. N. (2011). Shortest curves associated to a degenerate Jacobi metric on $\mathbb{T}^2$. In *Progress in variational methods*, Volume 7 of *Nankai Series in Pure, Applied Mathematics and Theoretical Physics*, pp. 126–168. World Scientific Publishing.

[66] Mather, J. N. and G. Forni (1994). Action minimizing orbits in hamiltomian systems. In *Transition to chaos in classical and quantum mechanics*, pp. 92–186. Springer.

[67] Mazzucchelli, M. and A. Sorrentino (2016). Remarks on the symplectic invariance of Aubry-Mather sets. *Comptes Rendus Mathématique. Académie des Sciences. Paris 354*(4), 419–423.

[68] McGehee, R. (1973). The stable manifold theorem via an isolating block. In *Symposium on Ordinary Differential Equations (Univ. Minnesota, Minneapolis, Minn., 1972; dedicated to Hugh L. Turrittin)*, Volume 312, pp. 135–144.

Springer-Verla.

[69] Oliveira, E. R. (2008). On generic properties of Lagrangians on surfaces: the Kupka-Smale theorem. *Discrete and Continuous Dynamical Systems 21*(2), 551–569.

[70] Pugh, C., M. Shub, and A. Wilkinson (1997). Hölder foliations. *Duke Mathematical Journal 86*(3), 517–546.

[71] Rifford, L. and R. O. Ruggiero (2011). Generic properties of closed orbits of Hamiltonian flows from mañé's viewpoint. *International Mathematics Research Notices 2012*(22), 5246–5265.

[72] Robinson, R. C. (1970). Generic properties of conservative systems. *American Journal of Mathematics 92*(3), 562–603.

[73] Shilnikov, L. P. (1967). On a Poincaré–Birkhoff problem. *Matematicheskii Sbornik 116*(3), 378–397.

[74] Sorrentino, A. (2015). *Action-minimizing methods in Hamiltonian dynamics (MN-50): An introduction to Aubry-Mather theory.* Princeton University Press.

[75] Treschev, D. (2002). Multidimensional symplectic separatrix maps. *Journal of Nonlinear Science 12*(1), 27–58.

[76] Treschev, D. (2004). Evolution of slow variables in a priori unstable Hamiltonian systems. *Nonlinearity 17*(5), 1803.

[77] Treschev, D. (2012). Arnold diffusion far from strong resonances in multidimensional a priori unstable Hamiltonian systems. *Nonlinearity 25*(9), 2717.

[78] Turaev, D. and L. Shilnikov (1989). On the hamiltonian-systems with homoclinic curves for a saddle. *Doklady Akademii Nauk USSR 304*(4), 811–814.

[79] Zhang, K. (2011). Speed of arnold diffusion for analytic Hamiltonian systems. *Inventiones Mathematicae 186*(2), 255–290.

[80] Zhang, K. (forthcoming). On the tangent cones of Aubry sets. *Annales de la Faculté des Sciences de Toulouse.*

Milton Keynes UK
Ingram Content Group UK Ltd.
UKHW051045250724
445942UK00015B/160